Praise for *We Have No Idea*

"Accessible and entertaining . . . Cham and Whiteson distill the essence of the little we know—and the lots we have no idea about. . . . A very enjoyable read."

—*Nature*

"This witty book reveals the humbling vastness of our ignorance about the universe, along with charming insights into what we actually do understand."

—**Carlo Rovelli, author of *Seven Brief Lessons on Physics* and *Reality Is Not What It Seems***

"[A] lucid and irreverent survey of the many unsolved mysteries of our universe . . . Cham and Whiteson mesh comics, lighthearted infographics, and lively explanations to painlessly introduce curious readers to complex concepts in easily digestible chapters. This fun guide is just the ticket for science fans of any age."

—*Publishers Weekly* (starred review)

"Cham and Whiteson perfectly balance hilarity and serious science."

—*Chemistry World*

"[A] lively, agnostic book on physics and its discontents . . . An entertaining and educational review for anyone seeking to brush up on or build his or her knowledge."

—*Kirkus Reviews*

"Science! Nerdy goodness! Cute illustrations! Big questions about the universe that we still can't answer! . . . Did I mention the cute illustrations?"

—*BookRiot*

"You couldn't ask for better guides to the mind-bending mysteries of cutting-edge physics than Jorge Cham and Daniel Whiteson. They bring a whimsical light touch to some very heavy topics, and the result is a sheer delight for the reader."

—**Jennifer Ouellette, author of *The Calculus Diaries* and *Me, Myself and Why: Searching for the Science of Self***

"Science knows a lot about the universe, but the real excitement is in what we don't know. And it's hard to imagine a more enjoyable way to explore the unknown than by reading this book. Jorge Cham and Daniel Whiteson will guide you through the biggest mysteries of the cosmos, smiling all the way."

—**Sean Carroll, author of *The Big Picture: On the Origins of Life, Meaning, and the Universe Itself***

"A delightful combination of comedy and cosmology that is as charming as it is informative."
—Zach Weinersmith, creator of *Saturday Morning Breakfast Cereal*

"Not often do you encounter such an optimistic and fun exploration of human ignorance."
—Henry Reich, creator of *MinutePhysics*

"Accessible and hilarious (the two best things it is possible for a book to be), *We Have No Idea* not only explores WHAT we don't know, but WHY we don't know it. You'd think that'd be plenty, but Cham and Whiteson also provide the most credible and up-to-date scientific explanations as to what some of the answers to these huge (and hugely important) questions might possibly be, PLUS puns."
—Ryan North, author of *Romeo and/or Juliet* and *To Be or Not To Be*

We Have No Idea

We Have No Idea

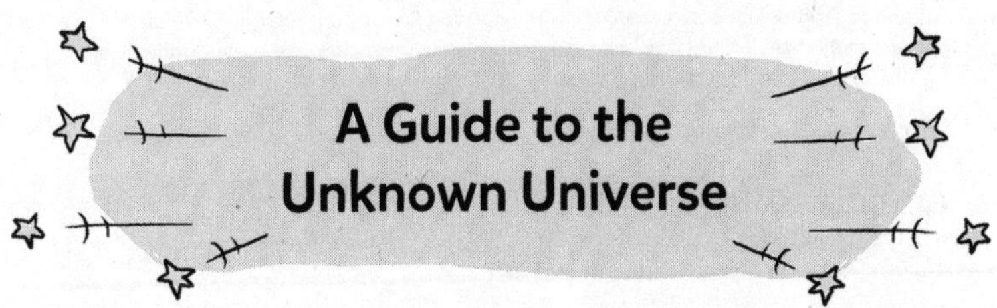

Jorge Cham and
Daniel Whiteson

RIVERHEAD BOOKS | NEW YORK

RIVERHEAD BOOKS
An imprint of Penguin Random House LLC
375 Hudson Street
New York, New York 10014

The Library of Congress has catalogued the Riverhead hardcover edition as follows:

Names: Cham, Jorge. | Whiteson, Daniel.
Title: We have no idea : a guide to the unknown universe / Jorge Cham and Daniel Whiteson.
Description: New York : Riverhead Books, 2017. | Includes bibliographical references and index.
Identifiers: LCCN 2016049070 | ISBN 9780735211513
Subjects: LCSH: Cosmology—Popular works.
Classification: LCC QB982 .C43 2017 | DDC 523.1—dc23
LC record available at https://lccn.loc.gov/2016049070
p. cm.

First Riverhead hardcover edition: May 2017
First Riverhead trade paperback edition: May 2018
Riverhead trade paperback ISBN: 9780735211520

Printed in the United States of America
5 7 9 10 8 6 4

Book design by Gretchen Achilles

For my daughter, Elinor.

—J.C.

To my family, for supporting all the chapters of my life, even those with bad puns in them.

—D.W.

Contents

We Have No Idea

Introduction

THE UNIVERSE AS WE KNOW IT:
EVERYTHING WE KNOW, EVERYTHING WE SEE, ALL THE ATOMS IN YOUR BODY AND IN OUR GALAXY, ALL THE STARS AND DUST AND PLANETS WITHIN AND OUTSIDE OF OUR SOLAR SYSTEM.
WE HAVE NO FREAKING IDEA.

Would you like to know how the universe began, what it's made of, and how it will end? To understand where time and space come from? To know whether we are alone in the universe?

Too bad! This book will not give you any of those answers.

Instead, this book is about all the things we *don't* know about the universe: all the big questions that you might think we have already answered but actually haven't.

We often hear on the news about some big discovery that answers a deep question about our universe. But how many people had heard of the question before they learned the answer? And how many big questions are still left unanswered? That's what this book is for, to introduce you to the open questions.

In the pages ahead, we'll explain what the biggest unanswered questions in the universe are and why they are still mysteries. By the end, you'll have a deeper grasp of just how absurd it is to think that we have any clue what's going on or how the universe really works. On the upside, at least you'll have a clue as to why we don't have a clue.

The point of the book is not to make you feel depressed about what we don't know but to fill you with a sense of excitement about the incredible amount of uncharted territory left to explore. For each unsolved cosmic mystery, we will also reveal what the answers could mean for humans and what mind-blowing surprises could be hiding in each unknown. We will teach you to look at the world in a different way—by understanding what we don't know, we can see that the future is still full of amazing possibilities.

So strap in, get comfortable, and get ready to explore the depths of our ignorance, because the first step in discovery is to know what is unknown. We are about to embark on a journey through the biggest mysteries in the universe.

1.

What Is the Universe Made Of?

In Which You Learn You Are Quite Weird and Special

ME CALL IT A "PARTICLE COLLIDER"

If you are a human being (we'll go with that assumption for now), then you probably can't help but be a little curious about the world around you. It's part of what it means to be human, and it's part of why you picked up this book.

It's not a new feeling. Since the dawn of time, people have wondered about the answers to some basic and very reasonable questions about the world around us:

What is the universe made of?

Are big rocks made of smaller rocks?

Why can't we eat rocks?
What is it like to be a bat?[1]

The first question, "What is the universe made of?" is a pretty big question. It's big not just because of the topic (it doesn't get much bigger than the universe), but because asking what the universe is made of is relevant to everyone. It's like asking what your house and everything in it (including you) are made out of. You don't need a deep understanding of mathematics or physics to understand that this question affects each and every one of us.

Say you were the first person to ever try to answer the question "What is the universe made of?" A good approach would be to try the simplest, most naïve idea first. For example, you might say that the universe is made of the things we can see in it, so you could answer the question by making a list. Such a list might start like this:

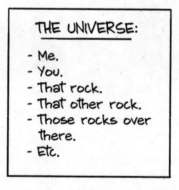

THE UNIVERSE:

- Me.
- You.
- That rock.
- That other rock.
- Those rocks over there.
- Etc.

But this approach has major problems. First, your list is going to be very, very long. It needs to include every rock on every planet in the universe, and it needs to include your list itself (it's also part of the universe). If you require that the list includes objects as well as the bits inside them,

1 The last question is the title of one of the most widely cited philosophy papers of all time, by American philosopher Thomas Nagel. Spoiler: the answer is "We can never know."

then it could be infinitely long. If you don't require the list to mention the bits inside objects on the list, then you could have a list of one element: "the universe." Clearly, this approach has big problems however you go about it.

But more important, making a list doesn't really answer the question. The kind of answer that would be satisfying wouldn't just record the complexity we see around us—the nearly infinite variety of stuff we see in our surroundings—it would *simplify* it for us, too. That is precisely the triumph of the periodic table of the elements (the one with oxygen, iron, carbon, etc.). It describes every object that humans have ever seen, touched, tasted,[2] or thrown at each other, all in terms of around a hundred basic building blocks. It reveals that the universe is organized under the same principle as Legos. With the same set of tiny plastic blocks, you can make toy dinosaurs, airplanes, or pirates—or create your own hybrid flying dino-pirate.

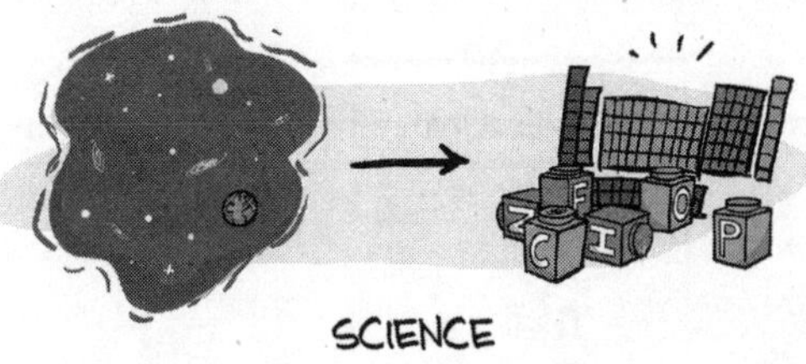

Just like Legos, a few basic building blocks (the elements) allow you to construct many things in our universe: stars, rocks, dust, ice cream, llamas. This organizing principle, where complex objects are really arrangements of simple objects, allows us to gain a deeper understanding by uncovering those simple objects.

But why does the universe follow the Lego philosophy? As far as we know, there is no reason why such a simplification is even possible. As far as the first cavemen and cavewomen scientists knew, the world *could* have worked in lots of different ways. All that cave scientists Ook and Groog

2 Yes, including that time in third grade your friend tasted a lizard.

had to base their ideas on was their experience, which was consistent with lots of different ideas about what the universe was made of.

It *could* have been that the number of kinds of stuff was nearly infinite. In such a universe, rocks could have been made out of elemental rock particles. Air could have been made out of elemental air particles. Elephants could have been made out of elemental elephant particles (let's call them Dumbotrons). In that hypothetical universe, the table of the elements would have a nearly *infinite number* of items.

EARLY PHYSICISTS

Or, even weirder, we could have lived in a universe where things are not made of tiny particles at all. In such a universe, rocks would just be made of smooth rock-stuff that can be cut into smaller and smaller pieces forever, and the knife you use to cut them would be infinitely sharp.

Both of those ideas were consistent with the data collected by Professors Ook and Groog in their famous rock-banging experiments. We mention these possibilities not because we think this is how the universe works but to remind you that it could have been how our part of the universe works, and *it might still be true for other kinds of matter in the universe that we have not yet explored.*

That's why the unanswered mysteries of the universe that you will discover in this book should make you feel inspired and excited rather than frustrated or demoralized. They reveal how much we have left to explore and discover.

In the universe we know and love, the things around us appear to be made out of tiny particles. After thousands of years of thought and research, we have a very fine theory of matter.[3] From Ook and Groog's first experiments to the modern day, we have surpassed the periodic table and peered inside the atom.

Matter as we know it is composed of atoms of the elements listed in

3 Science in its modern form with experiments and data and lab coats is only hundreds of years old, but the history of thought on these questions is thousands of years old.

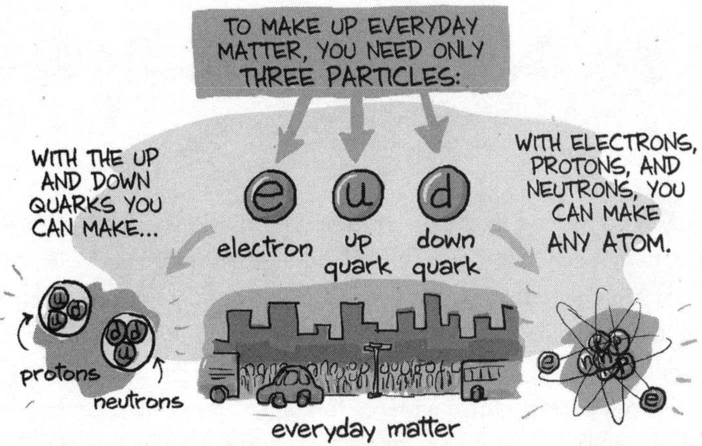
TO MAKE UP EVERYDAY MATTER, YOU NEED ONLY THREE PARTICLES:
WITH THE UP AND DOWN QUARKS YOU CAN MAKE...
WITH ELECTRONS, PROTONS, AND NEUTRONS, YOU CAN MAKE ANY ATOM.
e
u
d
electron
up quark
down quark
protons
neutrons
everyday matter

the periodic table. Each atom has a nucleus surrounded by a cloud of electrons. The nucleus contains protons and neutrons, each of which is built from up quarks and down quarks. So, with up quarks, down quarks, and electrons, we can build any element from the periodic table. What an achievement! We boiled down our list of the universe's ingredients from infinitely long to the hundred or so elements of the periodic table and then to only three particles. Everything we have ever seen, touched, smelled, or stubbed our toes on can be built from three basic building blocks. Congratulations to the collective work of millions of human brains.

But while we should feel proud of ourselves as a species, this description is incomplete in two very important ways.

First, there are other particles out there, not just the electron and two quarks. Only these three particles are needed to make normal matter, but in the past century, particle physicists have discovered nine more matter particles and five other particles that transmit forces. Some of these particles are very strange, such as the ghostly neutrino particles that can travel trillions of miles through lead without bouncing off of a single particle.[4] To neutrinos, lead is transparent. Other particles are very similar to the three particles that make up matter but are much, much heavier.

4 We think. Nobody has literally tried this experiment.

Why do we have these extra particles? What are they for? Who invited them to the party? How many other kinds of particles are there? We don't know. More than that: *we have no idea*. Some of these strange particles and their intriguing patterns will be discussed in detail in chapter 4.

But this description is incomplete in another very important way. While we need only three particles to build stars, planets, comets, and pickles, it turns out that these things make up only a tiny fraction of the universe. The kind of matter that we consider normal—because it's the only kind we know—is actually fairly unusual. Of all of the stuff (matter and energy) in the universe, this kind of matter accounts for only about 5 percent of the total.

What is the other 95 percent of the universe made of? *We don't know.*

If we drew a pie chart of the universe, it would look something like this:

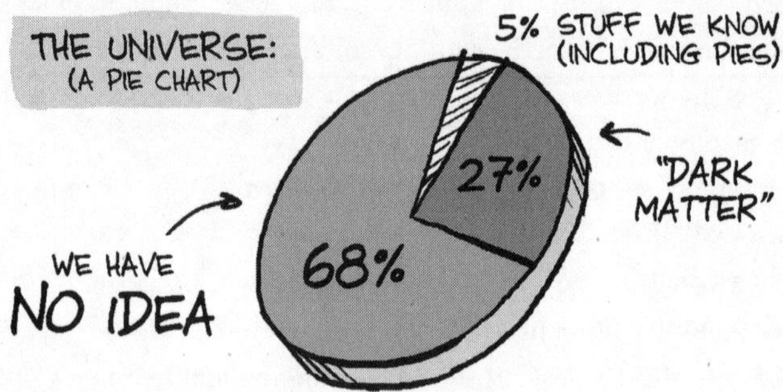

That pie looks pretty mysterious. Only 5 percent of it is stuff we know, including stars, planets, and everything on them. A full 27 percent is something we call "dark matter." The other 68 percent of the universe is something we barely understand at all. Physicists call it "dark energy," and we think it is causing the universe to expand, but that's about all we know about it. We'll explain both of these concepts and how we arrived at these exact figures in later chapters.

And it gets worse. Even within the 5 percent of stuff we know about, there are still a lot of things we don't know (remember those extra particles?). In some cases, we don't even know how to ask the right questions that will reveal these mysteries.

So this is where we stand as a species. Just a few paragraphs ago, we were congratulating ourselves on our incredible feats of intellectual exploration by describing all known matter in simple terms. Now that seems a bit premature, since *most of the universe is made of something else.* It's like we've been studying an elephant for thousands of years and suddenly we discovered we've been looking only *at its tail!*

Learning this, you might feel a bit disappointed. Maybe you thought we had reached the peak of our understanding and mastery of the universe (we have robots that will vacuum your house for you, for Pete's sake). But the important thing is to see this not as a disappointment but as an incredible opportunity: an opportunity to explore and learn and gain insight. What if you learned that we had explored only 5 percent of the land on Earth? Or that you had tasted only 5 percent of the world's ice cream flavors? The scientist in you would demand a thorough explanation (as well as more spoons) and be excited at the possibility of new discoveries.

Think back to elementary school when you were learning about the exploits of history's greatest explorers. They sailed into the unknown and discovered new lands and mapped the world. If you thought that sounded exciting, you might have also felt a twinge of sadness because now all the continents have been discovered, all the tiny islands have been named, and in this age of satellites and GPS, the era of exploration seems to be behind us. The good news is that this is not the case.

There's a *huge* amount of exploration left to do. In fact, we are in the early days of a whole new age of exploration. We are entering a period that will likely redefine our understanding of the universe. On one hand, we know that we know very little (5 percent, remember?), so we have some ideas of what questions to ask. And, on the other hand, we are building awesome new tools, such as powerful new particle colliders and gravitational-wave detectors and telescopes that will help us get the answers. This is all coming together *right now*.

The exciting thing is that the grand scientific mysteries *have* real, hard answers. We just don't know what they are yet. There is a chance that they could be solved in our lifetimes. For example, there either is or is not intelligent life somewhere in the universe right now at this very moment. The answer exists (Mulder was right: the truth *is* out there). Learning these answers would change at a very basic level the way we think about the world.

The history of science is one of revolutions in which we discover each time that our view of the world was distorted by our particular perspective. A flat Earth, an Earth-centered solar system, a universe dominated by stars and planets—these were all reasonable ideas given the data at the time, but we now see them as embarrassingly naïve. Almost certainly, there are more such revolutions around the corner, in which important ideas we accept now, such as relativity and quantum physics, might be shattered and replaced with mind-blowing new ones. Two hundred years from now, people will probably look back at our understanding of how things work the same way we look at how cavemen and cavewomen understood their world.

The journey of the human race to understand our universe is far from over, and *you* get to be a part of it. We promise that the ride will be sweeter than pie.

2.
What Is Dark Matter?

You're Swimming in It

Here is a bar chart of the mass and energy in the universe as we know it:

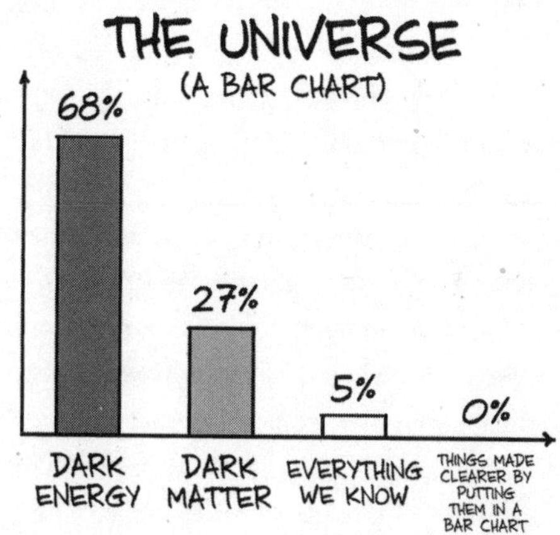

Physicists believe that an astounding 27 percent of the matter and energy in the known universe is made of something called "dark matter." This means that most of the matter in the universe is not the kind we have been studying for centuries. There is *five times* as much of this mysterious matter as there is normal, familiar matter. In fact, it's not really fair to call our matter "normal" when it's actually quite rare in the universe.

So what is this dark matter? Is it dangerous? Will it stain your clothes? How do we know it's there?

Dark matter is everywhere. In fact, you're probably swimming in it. Its existence was first proposed in the 1920s and first taken seriously in the 1960s when astronomers noticed something odd about how galaxies were spinning and what it meant for how much mass was inside them.

Ways We Know Dark Matter Is There

1. Spinning Galaxies

To understand the connection between dark matter and spinning galaxies, imagine putting a whole bunch of Ping-Pong balls in a merry-go-round. Now imagine giving it a spin. You would expect the Ping-Pong balls to fly off the edge of the merry-go-round. A spinning galaxy works almost the same way.[5] Because the galaxy is rotating, the stars in it tend to want to fly off outward. The only thing holding them together is the force of gravity from all the mass present in the galaxy (gravity pulls things with mass together). The faster the galaxy spins, the more mass you need to hold all the stars in. Conversely, knowing the mass of the galaxy means you can predict how fast the galaxy can spin.

Astronomers at first tried to guess the mass of galaxies by counting the number of stars in them. But when they used this number to compute how fast galaxies should be spinning, something didn't match up. Measurements showed that the galaxies were spinning faster than was predicted by how many stars they contained. In other words, the stars should be flying off the edges of the galaxies, just like the Ping-Pong balls in the merry-go-round. In order to explain the high rotation speed, astronomers needed to add a huge amount of mass to the galaxies in their calculation so all the stars held together. But they couldn't see where this mass was. This contradiction could be resolved if you assumed there was a huge amount of some kind of heavy stuff that was invisible, or "dark," in each galaxy.

This claim was quite extraordinary. And as the famous astronomer

5 Although galaxies tend to be slightly bigger than merry-go-rounds.

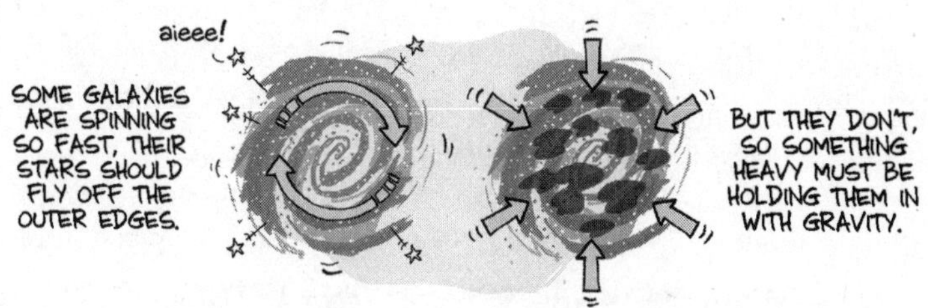

aieee!
SOME GALAXIES ARE SPINNING SO FAST, THEIR STARS SHOULD FLY OFF THE OUTER EDGES.
BUT THEY DON'T, SO SOMETHING HEAVY MUST BE HOLDING THEM IN WITH GRAVITY.

Carl Sagan once said, "Extraordinary claims require extraordinary evidence." So this strange conundrum existed in the astronomy community for decades without being understood. As the years went by, the existence of this mysterious invisible heavy stuff (or dark matter, as it became known) started to be more and more widely accepted.

2. Gravitational Lensing

Another important clue that convinced scientists that dark matter was real was the observation that it can *bend light*. This is called gravitational lensing.

Astronomers would sometimes look out into the sky and spot something strange. They would see the image of a galaxy coming from one direction. There's nothing weird about that, but if they moved the telescope a tiny bit, they would see the image of another galaxy that looked very similar to the first galaxy. The shape, the color, and the light that came from these galaxies were so similar, astronomers were sure they were the same galaxy. But how could this be? How could the same galaxy appear twice in the sky?

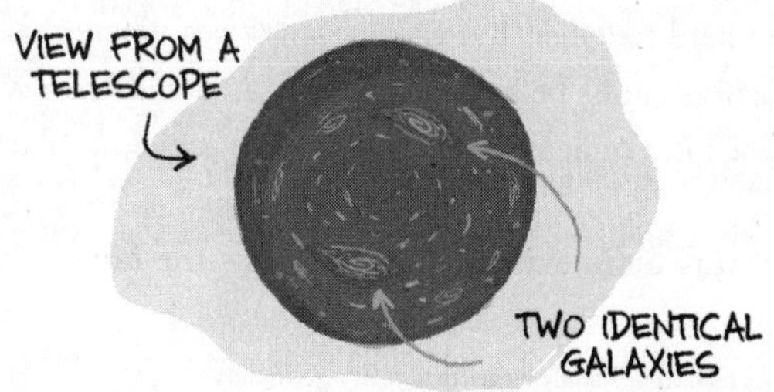

Seeing the same galaxy twice makes perfect sense if there is something heavy (and invisible) sitting between you and this galaxy; this invisible heavy blob can act like a giant lens, bending the light from the galaxy so it appears to be coming from two directions.

Imagine that light leaves this galaxy in all directions. Now picture two light particles, known as photons, coming from that galaxy and headed slightly to either side of you. If there is something heavy between you and that galaxy, the gravity from that object will distort the space around it, causing the light particles to curve toward you.[6]

On Earth, you see this in your telescope as two images of the same galaxy coming from different directions in the sky. This effect was observed all over the night sky; the heavy and invisible stuff seemed to be everywhere. Dark matter was soon no longer a crazy idea. There was evidence for it wherever we looked.

3. Colliding Galaxies

The most convincing single piece of evidence for dark matter came when we observed a giant galactic collision in space. Two clusters of galaxies crashed into each other millions of years ago in an epic event; we missed the collision itself, but since the light from it takes millions of years to reach us, we can sit back and comfortably watch the resulting explosions.

6 The bending of light due to gravity was something that Albert Einstein proposed and later proved. They say he was a pretty smart guy.

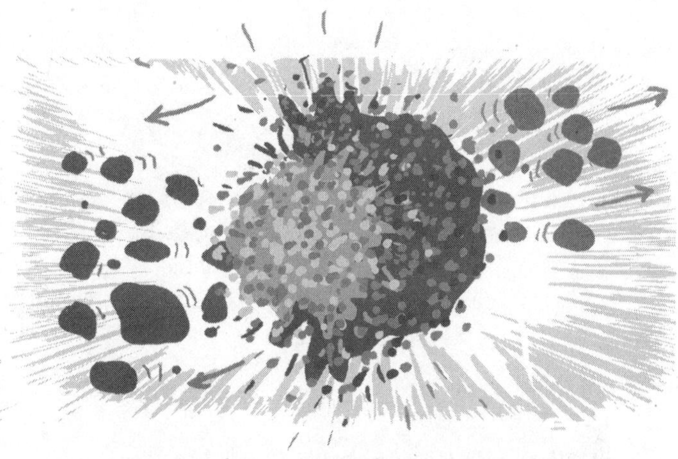

As the two galaxy clusters slammed into each other, the gas and dust from the two clusters collided with spectacular results: big explosions, giant clouds of dust getting ripped apart. It's a special-effects extravaganza. If it helps, visualize the collision of two huge piles of water balloons tossed at each other at crazy high speed.

But astronomers also noticed something else. Close to the collision site, they noticed two giant clusters of dark matter; of course, this dark matter was invisible, but they could spot it indirectly by measuring the distortion the clusters were causing to the light from the galaxies behind them. These two dark matter clusters seemed to be moving along the line of collision as if nothing had happened.

What astronomers have pieced together is this: there were two galaxy clusters, each with both regular matter (mostly gas and dust with some stars) and dark matter. When the two clusters collided, most of the gas and dust crashed together in the way you expect normal matter to do. But what happens when dark matter bumps into other dark matter? *Nothing that we could detect!* The clusters of dark matter kept going and passed *through* each other—almost as if they were invisible to each other. The stars also mostly passed through, because they were so sparse.

Enormous blobs of matter, bigger than many galaxies, passed right through each other. In essence, the collision stripped the gas and dust from these galaxies.

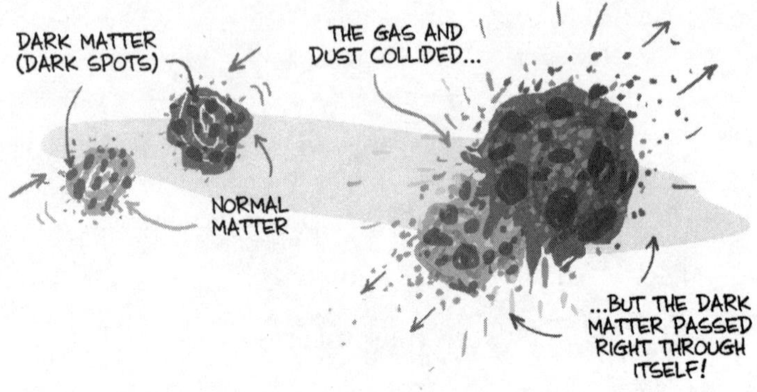
DARK MATTER
(DARK SPOTS)
THE GAS AND
DUST COLLIDED...
NORMAL
MATTER
...BUT THE DARK
MATTER PASSED
RIGHT THROUGH
ITSELF!

What We Know about Dark Matter

At this point, it should be pretty clear that dark matter exists and that it is something strange and different from the matter we are familiar with. Here's what we know about dark matter:

- It has mass.
- It's invisible.
- It likes to hang out with galaxies.
- Regular matter can't seem to touch it.
- Other dark matter can't seem to touch it either.[7]
- It has a cool name.

By now, you are probably thinking, *Man, I wish I were made of dark matter. I'd be an awesome superhero.* No? Okay, maybe that's just us.

One thing we know about dark matter is that it's not hiding far away. Dark matter tends to clump together in massive blobs that float in space and hang out with galaxies. That means there is a very high likelihood that dark matter surrounds you at this very moment. As you read this page, dark matter is very probably passing through this book and through

7 It's possible that dark matter can feel itself slightly through some new unknown force.

you. But if it is all around us, why is it such a mystery? Why can't you see it or touch it? How can something be there but not be seen?

It's hard to study dark matter because we can't interact with it very much. We can't see it (that's why it's called "dark"), but we know that it has mass (that's why it's called "matter"). To explain how this is all possible, we first have to think about how regular matter interacts.

How Matter Interacts

There are four major ways that matter interacts:

Gravity

If two things have mass, they will feel an attractive force toward each other.

Electromagnetism

This is the force that two particles feel if they have an electric charge. It can be attractive or repelling depending on whether the charges are different or the same.

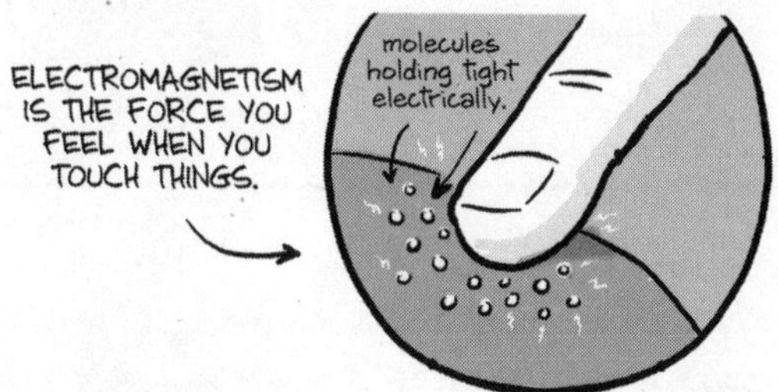

ELECTROMAGNETISM IS THE FORCE YOU FEEL WHEN YOU TOUCH THINGS.
molecules holding tight electrically.

You actually *feel* this force in your everyday life. If you press down on this book, the reason the paper doesn't get crushed, or the reason your hand doesn't go through the paper, is that the molecules inside the book are holding on tightly to one another with electromagnetic bonds and repelling the molecules in your hand.

Electromagnetism is also responsible for light and, of course, electricity and magnetism. We will talk more about light and the deep connections between particles and forces later on.

The Weak Nuclear Force

This force is similar in many ways to electromagnetism but is much, much weaker. For example, neutrinos use this force to interact (weakly!) with other particles. At very high energies, the weak force becomes as strong as electromagnetism and has been shown to be just one part of a unified force called "electroweak."

The Strong Nuclear Force

This is the force that keeps the protons and neutrons stuck together inside an atom's core. Without it, all those positively charged protons in the nucleus would simply repel one another and fly away.

How Dark Matter Interacts

It's important to note that this list of forces is only *descriptive*. Sometimes physics is like botany in that way. We don't understand *why* any of these forces exists. This is just a list of the things we've observed. We don't even know if this list is complete. But so far we can explain every experiment done in particle physics using these four forces.

So why is dark matter so dark? Well, dark matter has mass, so it feels gravity. But that's about all we know for certain about its interactions. We *think* that it doesn't have electromagnetic interactions. As far as we know, it doesn't reflect light or give off light, which is why it's hard for us to *see* it directly. Dark matter also doesn't seem to have weak or strong nuclear interactions.

So, barring any new undiscovered kind of interaction, it appears that dark matter cannot interact with us, or our telescopes or detectors, using any of the normal mechanisms. That makes it very hard to study.

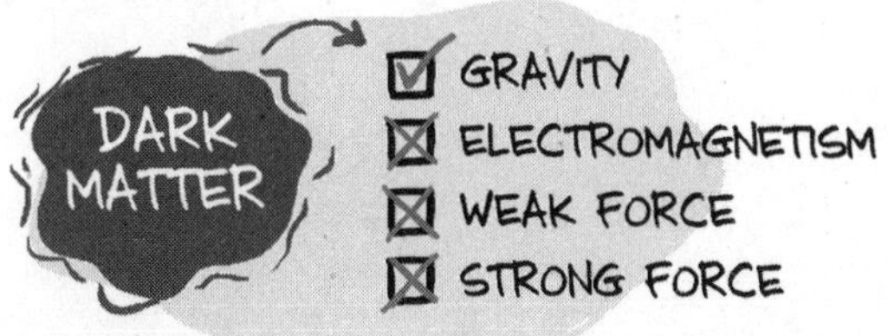

Of the four fundamental ways that we know things interact, the only one that we know for sure applies to dark matter is gravity. This is where the "matter" in dark matter comes from. Dark matter has stuff to it. It has mass, and if it has mass, it feels gravity.

How Can We Study Dark Matter?

We hope that we convinced you that dark matter exists. Something is definitely out there keeping stars from flying off into empty space, bending light from galaxies, and walking away from giant cosmic collisions the way action heroes walk away from car explosions in slow motion (without looking back). Dark matter is cool like that.

But the question remains: what is dark matter made of? We can't pretend to have an answer to the bigger question of what the universe is made of if we only study the easiest 5 percent. We can't ignore the whopping 27 percent that is dark matter. The short answer is that we still have very little idea what dark matter is. We know that it is there, how much of it there is, and roughly where it is, but we don't know what kind of particles it's made of—or even that it *is* made of particles. Remember that we need to be careful about extrapolating from one unusual kind of matter to the entire universe.[8] Keeping an open mind is necessary to make the kind of discoveries that change the way we think about the universe and our place in it.

To make progress, we need to examine some specific ideas, explore their consequences, and design experiments to test them. It's possible that dark matter is made of dancing cosmic purple elephants built out of a new and bizarre undetectable particle, but since that theory is difficult to test, it is not a top science priority.[9]

Sorry, you're
not a priority.

8 If you had a cheese sandwich for lunch today, that doesn't mean that all lunches are cheese sandwiches.

9 As of the date of this writing, science funding is unpredictable.

A simple and concrete idea is that dark matter is made of a new kind of particle that uses a new kind of force to interact very, very weakly with normal matter. Why consider only one new particle? Because it's the simplest idea, so it makes sense to tackle it first. It is definitely possible that dark matter is made of several kinds of particles like normal matter; these dark particles could have all sorts of interesting interactions, resulting in dark chemistry, perhaps even dark biology, dark life, and dark turkeys (a frightening thought).

This candidate particle is known by the acronym WIMP, which stands for Weakly Interacting Massive Particle (i.e., something with mass that interacts weakly with regular matter). We speculate that it might use a new hypothetical force to interact with our kind of matter at about the same level as neutrinos do, which is very, very little. For a while, people considered other ideas, such as really huge blobs of normal matter the size of Jupiter. To distinguish them from WIMPs, they were given the nickname MACHOs (Massive Astrophysical Compact Halo Objects).

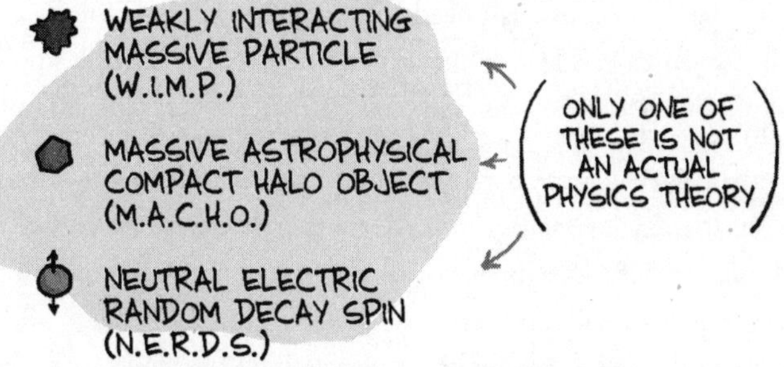

How do we know that dark matter particles interact with normal matter through other forces besides gravity? We don't. We hope they do, because that would make them much easier to detect. So we try the very difficult experiments before we try the almost impossible experiments.

Physicists have built experiments designed to detect these hypothetical dark matter particles. One classic strategy is to fill a container with a cold compressed noble gas and surround the container with detectors that go off when *one atom* of the gas gets bumped by dark matter. So far, these experiments have not seen any evidence of dark matter, but they are only now getting big enough and sensitive enough that we might expect them to detect dark matter.

Another approach is to try to create dark matter using a high-energy particle collider, which boosts normal matter particles (protons or electrons) to crazy high speeds and smashes them together. That's pretty awesome in and of itself, but it has the added benefit of being able to explore the universe for new particles. They have this power because they can turn one kind of matter into other kinds of matter. When particles smash together, they don't just rearrange the pieces inside them into new configurations; the old matter is annihilated and new forms of matter are made. It's like alchemy (we're not kidding) at a subatomic level. This means you can almost, with some limitations, make any kind of particle that can exist without knowing in advance what you are looking for. Scientists are examining the collisions to look for evidence that some of them lead to the creation of dark matter particles.

A third approach is to point our telescopes at places where we think there are high concentrations of dark matter. The closest one to us is the center of our galaxy, which seems to have a very large blob of dark matter. The idea is that two dark matter particles might randomly collide and annihilate each other. If dark matter has some way of interacting with itself, then dark matter particles could collide and turn into particles of normal matter, just as two normal matter particles can collide to create dark matter.[10] If this happens often enough, some of the resulting normal matter particles will have a particular distribution of energy and location that lets our telescopes identify them as likely to have come from dark matter collisions. But understanding this requires us to know a lot about what is happening at the center of the galaxy, which is another entirely separate set of mysteries.

10 If two normal matter particles can turn into two dark matter particles, then the process can also happen in reverse: two dark matter particles can turn into two normal matter particles.

Why This Matters

Dark matter is a big clue that for all of our discoveries and progress we are mostly still in the dark about the nature of the universe. In terms of our understanding, we are at the same level as cave scientists Ook and Groog. Dark matter is not even in our current mathematical or physical models of the universe. There is a large amount of stuff out there silently pulling on us, and we don't know what it is. We can't possibly claim to understand our universe without understanding this huge part of it.

Now, before you start feeling paranoid about weird, dark, mysterious stuff floating all around you, consider this: what if dark matter is something *awesome*?

Dark matter is made of something that we have no direct experience with. It's something we haven't seen before, and it might behave in ways we haven't imagined.

Think of the amazing potential that exists here.

What if dark matter is made of some new kind of particle that we are able to produce and harness in high-energy colliders? Or what if in discovering what it is, we figure out something about the laws of physics we didn't know about before, such as a new fundamental interaction or a new way that the existing interactions can work? And what if this new discovery lets us manipulate regular matter in new ways?

Imagine you've been playing a game your whole life, and suddenly you realize that there are special rules or special new pieces you could be playing with. What amazing technology or understanding could be unlocked by figuring out what dark matter is and how it works?

We can't stay in the dark about it forever. Just because it's dark doesn't mean it doesn't matter.

3.
What Is Dark Energy?

In Which Your Mind Is Exploded by Our Expanding Universe

You might be reeling from the fact that everything you thought you knew about the universe would barely get a 5 percent score on a standardized test administered by a race of smart star-traveling alien beings. Let's face it, your chances of attending alien university are probably pretty low.[11] To recap what we know as a human species, here is a stacked column chart of the universe (sorry, we're running out of chart types):

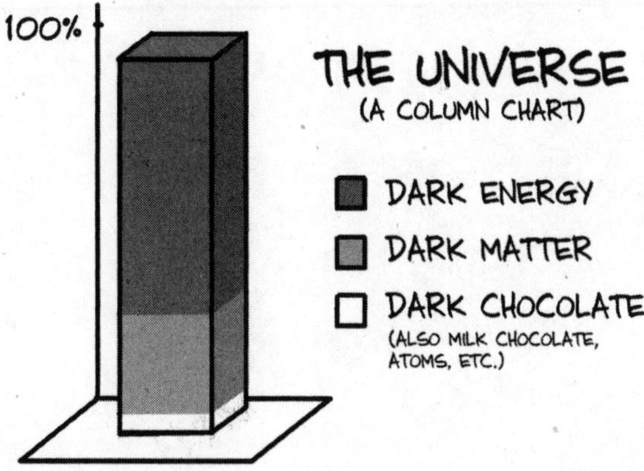

Imagine thinking all your life that you had an amazing and spacious house, and that it occupied your entire sense of everything there was. Then one day you discover it's actually only the bottom five floors in a

11 Which might be for the best—their cafeteria food is pretty weird.

one-hundred-story luxury apartment building. Suddenly, your living situation just got more complicated. Twenty-seven of those other floors belong to something heavy but invisible that we're calling dark matter. They might be cool neighbors or they might be weird neighbors. For some reason, they keep avoiding you in the hallways.

Fully sixty-eight of the other floors are nearly a *complete mystery*. This remaining 68 percent of the universe is what physicists are calling "dark energy." It's the biggest chunk of reality, and we have almost no idea what it is.

First, you might be wondering why it's called dark energy. The truth is that we could have called it anything.[12] Why anything? Because *we know almost nothing about it* except that it is causing the universe to *expand very rapidly*.

The second question you might have is "How do we know it's there?" And the answer is: quite by accident. It came as a total surprise to scientists, who were actually trying to answer a different question. They were trying to measure how quickly the expansion of the universe was slowing down, and instead they stumbled onto the fact that it wasn't slowing down at all but was expanding faster and faster. It's time to walk up the stairs and find out what these mysterious upper floors are all about.

THE UNIVERSE
IS EXPLODING.

Our Expanding Universe

To understand just how amazing and crazy it is that over two-thirds of the energy budget of the universe was discovered while looking for something

12 Well, almost anything. "The Dark Side" was taken.

else, we have to go back and start with the initial question that led to its discovery:

Does our universe have a beginning, or has it existed in its present form forever?

This might seem like a simple question, but it's actually quite profound. As recently as one hundred years ago, most sensible scientists thought it was *obvious* that the universe had been as it is for eternity and would continue to be that way forever. It had not even occurred to most people that our universe was changing. To them, all the stars and planets existed in a perpetual state of suspended motion, like a mobile hanging from the ceiling or a room full of clocks that never stop.

But then one day astronomers started noticing something odd. They measured the light from our surrounding stars and galaxies, and concluded that everything was moving apart from everything else. The universe wasn't just sitting there . . . it was *expanding*.

And if the universe had always been expanding, it meant that it's bigger now than it used to be. And if you continued to think this way and went back in time, you could imagine that the universe at some point was very small.

TRY NOT TO THINK
ABOUT WHOSE CEILING
WE'RE HANGING FROM.

Many physicists thought this was ridiculous and sarcastically called this theory the "Big Bang." If those scientists were living today, they would probably put their fingers up in the air, roll their eyes, and make ironic air quotes whenever they said it. It was a term meant to embarrass those who proposed this idea, but somehow it stuck. You know that

THE EARLY UNIVERSE

something's fundamentally changing our understanding of the universe when physicists start getting snarky.

So astronomers discovered in 1931 that the universe was expanding, which meant that it could be growing outward from an initial very, very[13] dense dot. (Note that this dot was not floating in some larger space, it *was* all of space. More on this crazy new way of thinking about space in chapter 7.) There were still some theories of a non–Big Bang universe consistent with the discovered expansion, but these theories required new matter to be constantly created to keep the expanding universe at the current density.

If the universe had a beginning, then it makes you immediately wonder about whether it will have an end. What could possibly bring this enormous, majestic, and wonderfully strange place to an end? And most important, do you have time to finish that novel you have been working on forever?

What could possibly cause the universe to end? The answer is our old friend gravity.

Remember that while all the stuff in the universe is shooting out from the cosmic explosion of the Big Bang, gravity is working in the other direction. Every bit of matter in the universe feels gravity, which is doing its best to pull the universe back together. What does that mean for the eventual fate of the universe? People had several ideas (see the next page).

Here's the mind-blowing part. The actual answer is: *none of these*! The truth, as strange as it might be, is a *secret* fourth option that only a few scientists considered (because it seemed totally crazy):

13 We can't write enough *verys* here to communicate how dense this dot was. It was the entire universe condensed into *one point*.

POSSIBLE FATES OF THE UNIVERSE	CORRESPONDING EMOJI
A. THERE IS SO MUCH STUFF IN THE UNIVERSE THAT THE FORCE OF GRAVITY WILL EVENTUALLY WIN, SLOW DOWN THE EXPANSION, AND SHRINK EVERYTHING BACK DOWN. THIS IS CALLED THE BIG CRUNCH.	:O
B. THERE IS NOT ENOUGH STUFF IN THE UNIVERSE FOR GRAVITY TO SLOW DOWN THE EXPANSION, SO THE UNIVERSE KEEPS EXPANDING FOREVER UNTIL IT SPREADS OUT INTO AN INFINITELY DILUTE AND COLD (AND LONELY) UNIVERSE.	:(
C. THERE IS JUST ENOUGH STUFF FOR GRAVITY TO SLOW DOWN THE EXPANSION, BUT NOT QUITE ENOUGH TO STOP IT OR SHRINK THE UNIVERSE BACK DOWN. THE UNIVERSE KEEPS EXPANDING, BUT THE EXPANSION SLOWLY APPROACHES ZERO.	:|

Some incredibly powerful and mysterious force is expanding space itself so the universe is growing faster and faster.

This fourth option is the only one consistent with what we observe about our universe.

THE END IS FAR (AND GETTING FARTHER)
I BELIEVE IN SCIENTIFICALLY ACCURATE DOOMSAYING.

How We Know the Universe Is Expanding

This question of the fate of the universe seems like a very important one, but you can relax. The future we are discussing is billions and billions of years away regardless of what happens. You have time to finish your best-selling novel and even write a sequel. But this topic is important to us because, when we find answers to big questions like these, we also understand more about how our universe works. Sometimes, in asking these questions, we learn something surprising that can affect our day-to-day existence. For example, do you appreciate the GPS feature on your phone? An accurate GPS system is possible only because Einstein had questions about what happens when things move at the speed of light, which doesn't happen often here on Earth. But this led to the development of relativity, without which GPS would not be accurate.

To predict the eventual fate of the universe, scientists needed to know how quickly the universe is expanding. They did this by measuring the speed at which the galaxies around us are moving away from us.

First, you should understand that in an expanding universe everything is moving away from everything else, not just away from the center. Imagine we are a raisin in a universe-size loaf of raisin bread. As the bread bakes and rises, all the raisins move away from all the other raisins, but the raisins stay the same size.

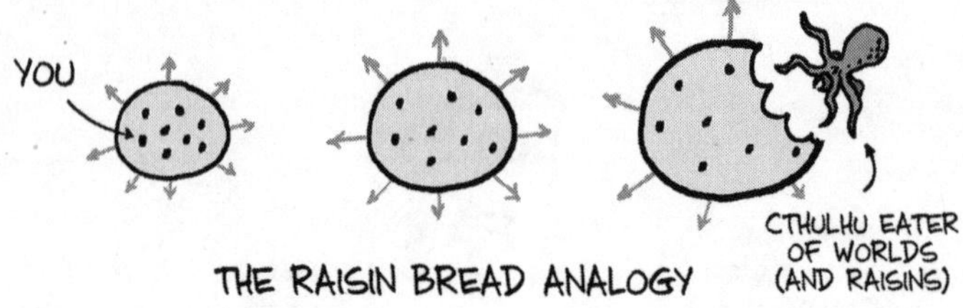

THE RAISIN BREAD ANALOGY

But to know the fate of the universe, we want to know whether this expansion is changing: are other galaxies moving away from us more *slowly* now than they were a few billion years ago? Or are they moving away from us more *quickly* than they were a few billion years ago? What we want to know is how the rate of expansion is *changing over time*. To see this, we need to know how fast things were moving away from us in the *past* and compare that with how fast things are moving away from us *now*.

Seeing the future is very difficult, but for astronomers, looking into the past is easy. Since the universe is so enormous and light has a finite speed, it takes the light from distant objects a long time to reach the Earth. This means that light from stars very far away is *very old light*, and the information it carries is also old. Looking at this light is like looking backward in time.

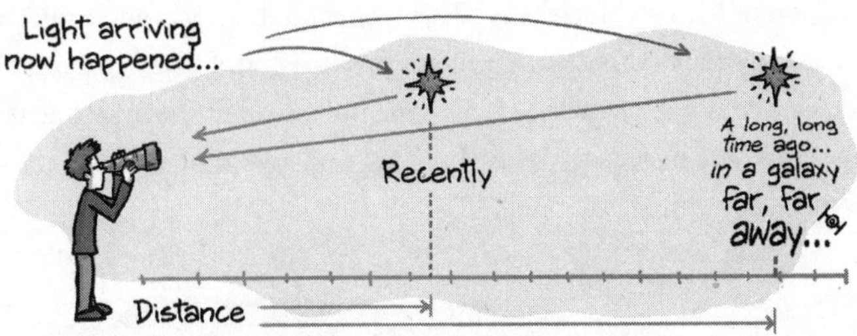

And it works the other way, too. If aliens on a planet that is really far away are looking at the Earth through their telescopes, they see light that left the Earth a long time ago. Right now, they could be watching that extremely embarrassing incident that happened to you years ago (*you know which one*).

So, the farther away an object is, the older the light we see and the further back in time we can look. This means that if we see faraway objects moving at one speed, and then we see closer objects moving at another speed, we can deduce that the speed of things has changed with time. We can measure the speed of a distant star from the shift in the frequency spectrum of its light using the same technique (the Doppler effect) that the police use to give you speeding tickets. The faster a star is moving away from us, the redder its light will be.

Knowing how far away things are required some clever sciencing.[14] For example, how do you tell the difference between a dim star that's close by and a bright star that's far away? Through a telescope, they look the same: like little dim points of light in the night. That was true until

14 That's right, we used it as a verb.

scientists identified a special kind of star, one that very predictably did the same thing everywhere in the universe. Because of their size and composition, these special stars grow at the same rate, and when they reach a certain size, they always do the same thing: they explode. Or to be more accurate, they implode, but the implosion is so violent, it generates a corresponding big explosion.[15] This type of explosion is called a type Ia supernova. What's useful about these supernovae is that, generally speaking, they all explode in a similar way. This means that, after some calibration, if you see one that's dim, you know it's far away, and if you see one that's bright, you know it's nearby. It's like the universe put these identical beacons everywhere just so we know how big and awesome it is (the universe is mysterious but not humble).

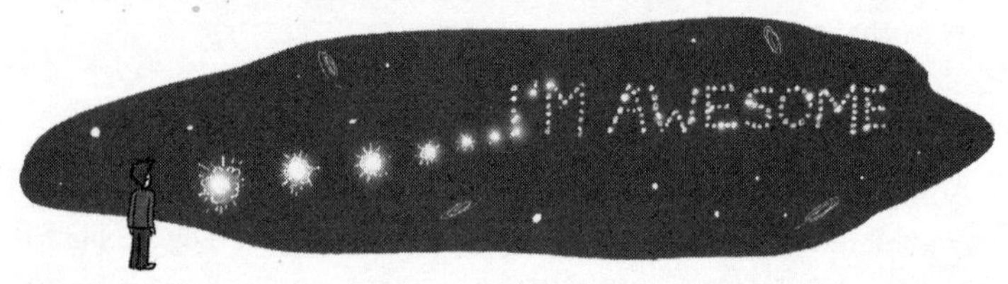

Astronomers call these type Ia supernovae "standard candles" (they're romantic like that). With them, astronomers could tell how far away (and therefore how old) distant objects were, and using the Doppler shift, they could tell how fast they were going. This meant astronomers could measure how the expansion of the universe was changing.

Shortly after realizing this, two teams of scientists raced against each other to measure the rate of expansion of the universe. But finding supernovae is not easy because they are short-lived explosions. To catch one, you have to constantly scan the sky for stars and spot the ones that suddenly get much brighter and then dimmer so it took a while.

The two teams assumed that the expansion of the universe should either be slowing down or staying the same. This is a reasonable assumption. If the universe exploded, and gravity is trying to pull everything back

15 Astronomy: more explosions than a Michael Bay movie.

in, then there are only two options: either gravity wins and things get pulled back in, or it loses and everything keeps expanding steadily.

When the scientists measured these supernovae and calculated the rate at which the universe was expanding, they expected gravity to be winning. That is, they expected to find that more distant stars (the ones in the past) were moving away more quickly than closer stars (the ones closer to the present). Instead, they were flummoxed to discover the opposite: that stars seem to be moving away from us more quickly now than they were in the past. In other words, the universe is expanding *faster* now than it was before.

Let's take a moment to consider just how unexpected this result was. In the astronomers' minds, there were two things: a universe that exploded a long time ago and gravity, which is trying to pull it all together again. Instead, there is a critical third piece: the size of space itself. As we will discuss in gory detail in chapter 7, space is not a static empty backdrop on which the theater of the universe plays out. It is a physical thing that can bend (in the presence of massive objects), ripple (called gravitational waves), or expand. And it appears that it is expanding—and quickly. Space is *rushing* to get bigger. Something is creating more space, which pushes everything in the universe outward.

We should note that the actual results showed that things *were* slowing down at first, but for the last five billion years, something has been pushing the bits of the exploding universe faster and faster away from one another.

This driving force that's making the universe bigger at an increasing rate is what physicists call dark energy. We can't see it (that's why it's "dark"), and it's pushing everything apart (so they call it an "energy"). And it's such a major force that it's estimated to represent 68 percent of the total mass and energy in the universe.

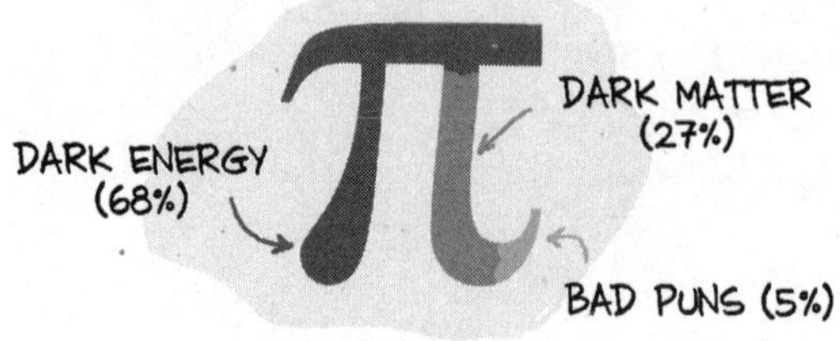

The Pie Chart

Up to now, we've been very specific in our labeling of our universe pie chart. Five percent sounds like an estimate, but when you hear percentages like 27 for dark matter and 68 for dark energy, you have to imagine that physicists are using more than a wild guess to come up with these figures.

So how do we know how much dark matter and dark energy there is in the universe?

For dark matter, we can't measure all the bits of it using the tools we learned about before (gravitational lensing and spinning galaxies) and add it all up. There isn't always the right arrangement of stars and dark matter to use these methods, and there could always be more dark matter hiding somewhere where we can't find it.[16]

And for dark energy, we don't really know what it is, so we can't measure it directly either.

The impressive part, given our lack of understanding of what these things are, is that we have managed to measure these percentages in several different ways. And so far, all of them seem to agree.

The most precise way we know how much dark matter and dark

16 With all your lost socks and misplaced keys.

energy there is comes from examining a baby picture of the universe: a photograph of the universe when it was still tiny and cute.[17]

We'll talk in later chapters about how this baby picture of the universe was taken and what it represents, but for now, just know that such a picture exists. This picture is called the cosmic microwave background and it looks something like this:

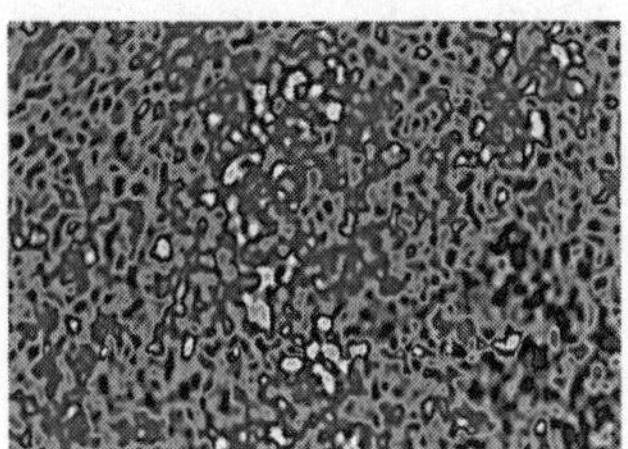

THE BABY UNIVERSE
(DIAPERS NOT INCLUDED)

Okay, it's not that cute. In fact, it's kind of a lumpy mess covered in wrinkles (like most babies). This picture captures the first photons that escaped the early formation of the universe. What's important is that the number of wrinkles and the patterns they form in the picture are very sensitive to the proportion of dark matter, dark energy, and regular matter in the universe. In other words, if you change the proportions, then the patterns in the picture will come out differently. It turns out that for the patterns that we see in the picture you'd need about 5 percent regular matter, 27 percent dark matter, and 68 percent dark energy. Anything else would give us a different picture than what we observe.

Another way we've measured dark energy is by looking at the rate of expansion of the universe, which we know from the supernova standard candles. We know dark energy is pushing everything outward at a faster and faster speed. From our estimates of matter and dark matter, we can calculate how much dark energy would be needed to get that expansion, and that gives us an estimate of the amount of dark energy there is.

17 It's always good to flatter the thing that made you.

And finally, we can tell the proportions of dark matter, dark energy, and regular matter by looking at the structure of the universe that we see today. The universe is arranged in a very particular configuration of stars and galaxies. Using a computer simulation, we can backtrack from this present state to just after the Big Bang to see how much dark matter and dark energy you'd need to get things to look the way they are now. For example, if you don't have the right amount of dark matter in the simulation, then you don't get galaxies in the same shape as we see them now, and they don't form as early as we know they did. Dark matter, because of its enormous mass and gravitational pull, helps normal matter clump together in the way that's needed for galaxies to form early. At the same time, if you try to explain all the energy in the universe in terms of only matter and dark matter, with no dark energy (i.e., dark matter = 95 percent), then the galaxies don't come out correctly either.

What's amazing is that all of these methods agree with one another.

THE PIE DON'T LIE

They all reveal that our universe is roughly made up of a combination of regular matter, dark matter, and dark energy that's 5 percent, 27 percent, and 68 percent. Even though we don't know what each of these things are, we can say with fairly good confidence that we know how much of them there is. We have no idea what they are, but we know they are *there*. Welcome to the era of precision ignorance.

What Could Dark Energy Be?

We have shown you how dark energy was discovered and how much of it there is, but what is it? The short answer is that *we have no idea*. We

YOU'RE MY
FATHER??
I SAID, "I'M
MUCH FARTHER"

NEVER UNDERESTIMATE THE POWER OF THE DARK ~~SIDE.~~ ENERGY EXPANSION.

know that it's a force that's currently *expanding* the universe. It's taking everything that's matter in the universe and pushing it outward. Right now, it's pushing me, it's pushing you, and it's pushing everything we know away from one another.[18] And we don't know what it is.

One currently popular idea is that dark energy comes from the energy of empty space. Yes, *empty space*.

When we say something is empty, we mean that it has no "stuff" inside of it. To be more technical, we think of it as having no stuff *to* it. There are places in intergalactic space that simply have no matter particles (not even dark matter). Now consider this: what if this empty space had energy to it, like a glow or a low hum, even if it has no matter? It just has energy that sits there for no good reason. If this was true, that energy could provide a gravitational effect that pushes the universe outward.

EMPTY SPACE ENERGY
(TRUST US, IT'S THERE)

This may sound crazy, but it's actually a surprisingly reasonable

18 It's not love that'll tear us apart; it's dark energy.

explanation. In fact, it is quite natural in quantum mechanics to have a vacuum energy. According to quantum mechanics, the world works very differently for very small objects (like particles) than it does for larger objects (like people and pickles). Quantum objects can do things that make little sense for pickles to do, like not having a precisely defined location, appearing on the other side of impenetrable barriers, and acting differently depending on whether they are being observed. Also, according to quantum physics, particles can pop into existence and back out again from the energy of otherwise empty space.

After all, it was quantum mechanics that gave us a different view of reality, and relativity that made us abandon the idea of absolute space or time. So why not accept that what appears to be empty space is full of vacuum energy pushing the universe apart?

One problem with this theory is that when scientists try to calculate how much energy empty space should have according to quantum mechanics they get an answer that is too big. And not just a little bit too big, but 10^{60} to 10^{100} times too big. That's a googol too big (google it). For comparison, the estimate of the number of particles in the entire universe is only 10^{85}. So it is fair to say that this idea would overshoot it a bit.

FIELD OF ENERGY DREAMS

Other ideas include new forces or special fields that permeate space just as the electromagnetic field does. Some of these fields are conceptualized to vary with time to explain why the accelerating expansion of the universe began only five billion years ago. There are a lot of different versions of these theories, but the thing they have in common is that they are difficult to test. After all, some of these fields might not interact with our particles, making it hard to design an experiment to detect them. Some of the fields might also feature new particles (like the Higgs field has the Higgs boson), but those particles could be very, very massive, making them way out of the range of what we can measure today. How massive? Heavier than anything we have seen before but not as heavy as your cat.

All of these ideas are in their infancy. They are just the initial proto-ideas that will lead scientists to better ideas until eventually we understand what most of the energy in the universe is up to. By comparison, dark energy makes dark matter look very simple and well understood: at least we know that it is matter. Dark energy could almost literally be anything. If a scientist from five hundred years in the future looked back in time at us, our current ideas about dark energy might seem hilarious to her, the way early men and women explaining the stars, the Sun, or the weather as being the result of gods dressed in robes seems quaint to us now. We know that there are powerful forces out there beyond our comprehension and that we have much to learn about the universe.

What This Means about the Future

If the universe is expanding more and more quickly because of dark energy, it means that everything is getting farther away from us a little faster each day. As the expansion picks up speed, things that are far apart will eventually be expanding away from one another *faster than the speed of light*. This means that light from stars will stop being able to reach us. Already, there are fewer stars visible to us in our night sky today than there were yesterday. If you follow this expansion to its natural conclusion, in billions of years the night sky will have only a few visible stars. And, even further into the future, the night sky could be almost totally dark.

Imagine that you were a scientist on that future Earth. How would you guess at the existence of stars and galaxies that you can't see?[19] If the expansion continues, it could eventually rip apart our solar system, our planet, even the smartphones out of your great^{n}-grandchildren's hands. On the other hand, since we know so little about what is driving this expansion, it might also be the case that it slows down in the future.

But it makes you think: if there were once more stars visible to us than there are today, what once-obvious facts are we missing because humans arrived nearly fourteen billion years after the party started?

THE FUTURE NIGHT SKY :/

19 If you want to see the stars, better not put off that camping trip another billion years.

4.

What Is the Most Basic Element of Matter?

In Which You See How Little We Understand about the Littlest Bits We See

Learning that all of human knowledge and science is relevant only to the 5 percent of the universe that we call "normal matter" can lead to several possible reactions. It might:

- **a.** Make you feel small, humbled, and slightly terrified.
- **b.** Cause you to deny, deny, deny.
- **c.** Arouse your excitement for all the things we can learn about the universe.
- **d.** Encourage you to keep reading this book.[20]

If your reaction is to feel humbled and terrified, we have good news for you. We are going to spend most of this chapter talking about normal matter. By the way, if dark matter does have dark physics, dark chemistry, dark biology, and, by extension, physicists made of dark matter, they would probably argue that *their* matter is "normal." Maybe you *should* feel a little humbled.

We also have bad news for you. We don't know everything there is to know about the 5 percent we know something about.

This may come as a surprise to a lot of you. After all, we have been around for only a few hundred thousand years, and we've done pretty

20 And buy copies for all your friends.

well for ourselves in terms of science. In fact, you might be tempted to say that we've mastered our little corner of the universe. We have so much snazzy technology at our fingertips today, you'd think we have a pretty good handle on the science of everyday matter. We can stream hours of bad TV shows wherever and whenever. Surely that's a milestone in any civilization.

Interestingly enough, this is both true and false (the idea that we have a good handle on reality, not that we can watch reality TV on our screens 24/7).

It is true that we know a lot about everyday matter. But it's also true that there's a lot we *don't* know about everyday matter. Most notably, we have no idea what some particles (bits of matter) are even for. Here's where we stand: in the everyday business of physics exploration, we've discovered twelve matter particles. Six of those we call "quarks" and the other six we call "leptons."

Yet you only need three of those twelve to make up everything around you: the up quark, the down quark, and the electron (one of the leptons). Remember that with the up and down quarks, you can make protons and neutrons, and together with the electron, you can make any atom. So what are the other nine particles for? Why are they there? *We have no idea.*

How puzzling is this? Well, imagine you made this great cake and after baking it and decorating it, and tasting it (it tastes great, by the way; you are an excellent baker), you discovered you had nine other ingredients you didn't even use. Who put those ingredients there? Were they supposed to be used for something? Who came up with this recipe anyway?

The truth is that our lack of knowledge about everyday matter (the 5 percent) goes a lot deeper than particle pastry.

THE CAKE OF SCIENCE

To recap, we understand how three particles (up quarks, down quarks, and electrons) can be combined to make any kind of atom. And we know how atoms can be used to make molecules, and how molecules can form complex objects like cakes and elephants. But all of that is just the *how*: we know how things go together, and we know how to put them together. We know this so well that we can make everything from sweat-wicking underwear to space telescopes. We are pretty amazing, right?[21]

What we don't know very much about is the *why*: Why are things put together the way they are? Why aren't they put together in a different way? Is this the only version of a self-consistent universe, or are there 10^{500} different versions as proposed by string theorists?

We don't know yet at a fundamental level the reason all the pieces in the universe fit together. It's like music: we know how to make music, we dance to it, everybody sings along to it, but we don't know why it works to get us grooving. It's the same with the universe: we know it works, but we don't know *why* it works.

Some might argue that such an explanation doesn't exist, or that if it

21 And yet, still no flying cars.

does exist we may never be able to know it, much less comprehend it. We'll leave that discussion for chapter 16, but the point is that we definitely don't have that knowledge today.

Now, assuming you are a curious person and are genuinely interested in knowing the why of things,[22] you might be wondering how to go about answering this question, and what it has to do with the useless particles we've found.

Well, if we're going to understand the basic "why" of the universe, the first thing we need to do is figure out what the universe is like at its deepest, most fundamental level. This means breaking the universe down until we can't break it down anymore. What is the smallest, most basic bit of reality? If that bit is a particle, then we want to find the particles that make up the particles that make up the particles that make up the particles, etc., ad infinitum (or ad nauseam, whichever comes first).

Once you find such elemental particles, you could then examine them and possibly figure out why everything works the way it does. It would be like finding the smallest Lego pieces in a Lego universe. If you found those, you would know what the basic system is for how everything interlocks with everything else. You would know something deep and true about reality, including (we hope) dark energy and dark matter.

Right now, we're not sure if we know the universe down to its smallest possible size. Or if we do, we're not sure what to make of the Lego pieces we've found. But the exciting thing is that we have a map. We have an incomplete crossword puzzle of the universe, and this crossword puzzle looks a lot like something we've seen before: it looks like a periodic table.

22 This is a reasonable assumption, given that you read even the footnotes.

The Periodic Table of Fundamental Particles

After a century of smashing things around, physicists have found that the twelve fundamental matter particles can be arranged in a table that looks something like this:

THE "FUNDAMENTAL" MATTER PARTICLES

	1st GENERATION	2nd GENERATION	3rd GENERATION	CHARGE
QUARKS:	UP u	CHARM c	TOP t	+2/3
	DOWN d	STRANGE s	BOTTOM b	-1/3
LEPTONS:	ELECTRON e	MUON μ	TAU τ	-1
	$\nu_{ELECTRON}$ ν_e	ν_{MUON} ν_μ	ν_{TAU} ν_τ	0
	SOME MASS	MORE MASS	EVEN MORE MASS	

Let's take a moment to appreciate how significant it is that we've gotten to this point. Remember that cavemen physicists Ook and Groog's initial theory of the universe was:[23]

THEORY OF THE UNIVERSE:
By Ook and Groog

THE UNIVERSE IS:

- Ook and Groog.
- Ook's favorite rock.
- Groog's pet llama.
- Yadda Yadda Yadda.

23 This "yadda yadda yadda" holds the record for largest amount of matter yadda-yadda-yadda'd.

This was a complete picture, but it wasn't helpful because it didn't tell us anything fundamental or insightful; it's a statement of the obvious. Later on, the Greeks had the idea that everything was made of four elements: *water, earth, air, and fire*. This was flat-out wrong, but at least it was a step in the right direction because it tried to *simplify* the description of the world.

Then we discovered the elements and that rocks and earth and water and llamas are all made of a small set of different kinds of atoms. Later, we found out that even atoms are made of smaller particles, and some of those are made of even smaller particles (quarks). The most important lesson we've gained from all this is that atoms and llamas are not the elemental units of the universe. If there is a fundamental equation of the universe—whatever it is—we can be sure it doesn't have a variable called N_{llamas} because llamas, like atoms, are not a fundamental element of the universe. They don't define its essential nature; they are just the aggregate result (the emergent phenomenon) of the deeper reality (sorry, llamas) the same way tornados are an emergent phenomenon of wind or stars are an emergent phenomenon of gas and gravity.

NOT THE FUNDAMENTAL UNITS OF THE UNIVERSE:

ATOMS

LLAMAS

TORNADOES

LLAMA-TORNADOES

Organizing what we know (and don't know) into tables helps us notice if there are patterns and missing pieces. Imagine for a moment that you were a scientist in the 1800s (yes, you can imagine wearing silly spectacles), and you didn't know yet that atoms are actually made of smaller electrons, protons, and neutrons. If you organized what you did know into a periodic table of the elements, you would have noticed some interesting things.

You would have noticed that the elements on one side of the periodic

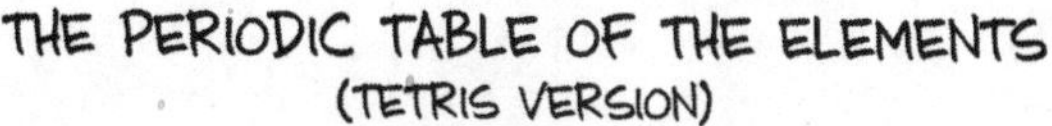

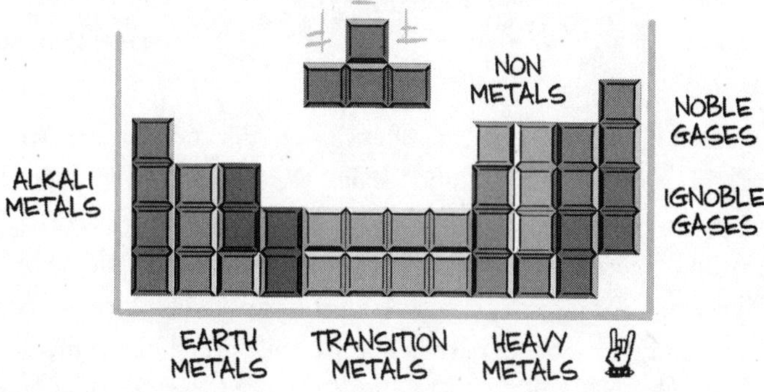

table are very reactive while the ones on the other side are almost totally inert—and that groups of nearby elements have similar properties, such as the metals, and that some elements are harder to find than others.

All of these curious patterns would have given you clues that the periodic table was *not* the fundamental description of the universe. They imply that something deeper is going on. It's like meeting a group of siblings and noticing certain similarities among them. Even though they're all different, you might assume they came from the same two parents because of the way they look or act. In the same way, scientists looked at early versions of the periodic table, noticed the patterns, and wondered, *Are we missing something?*

Now we know that the patterns in the periodic table are due to the arrangement of electron orbitals, and we know that there is an element for every spot and that some elements are rarer than others because they decay radioactively. It's all just a matter of putting together the right number of neutrons, protons, and electrons to get every element.

The point is that we organized the knowledge that we had at the time and we studied it carefully. Then we started to notice patterns and missing pieces, and this led us to ask the right questions, which led us to have a deeper understanding of how the universe works.

It took most of the twentieth century to put together that table of

HOW TO SCIENCE:
ORGANIZE WHAT YOU KNOW
LOOK FOR PATTERNS
ASK QUESTIONS
BUY TWEED JACKET WITH ELBOW PATCHES

fundamental matter particles (the one with quarks and leptons). We call these particles "fundamental" not because they are fun (they totally are) but because we can't yet see if they are made of even smaller particles. We actually don't have any proof that they are the most basic building blocks in the universe, but they're the smallest bits of stuff we've seen (so far).

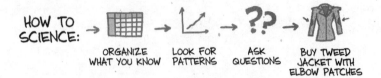

If you study the table of particles on page 47, you'll notice that it has some interesting patterns, too. First, you'll notice there are two kinds of matter particles: quarks and leptons. We know they're different because quarks feel the strong nuclear force, but leptons do not. Then you might notice that the particles that make up everyday matter are all in the first column: the *up quark*, *down quark*, and *electron*. There's a fourth particle in that first column called the electron neutrino (ν_e), and it speeds through the cosmos like a ghost, not really interacting with anything much at all.

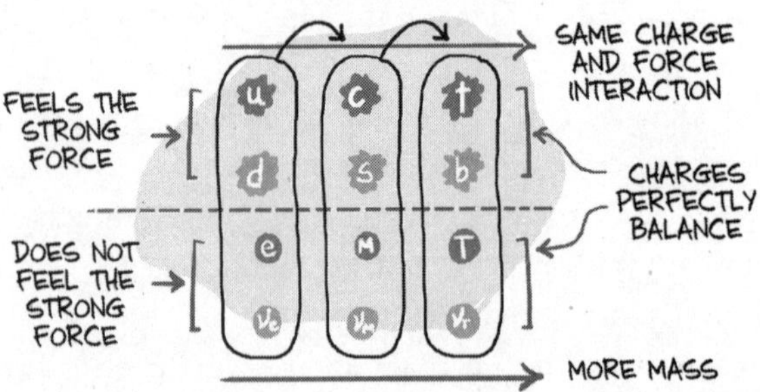
PARTICLE PATTERNS
SAME CHARGE AND FORCE INTERACTION
FEELS THE STRONG FORCE
CHARGES PERFECTLY BALANCE
DOES NOT FEEL THE STRONG FORCE
MORE MASS

But wait, there's more! There are other particles besides these four and they all fall into columns as well. Each column looks exactly like the first column (with the same properties like charge and force interaction) except the particles in them have more mass.[24] We call each of these columns a "generation," and we've discovered three of these generations.

You might immediately have some questions about our table of particles:

- Does it come in birch?
- What are all these particles for?
- What's the pattern of the masses of the particles?
- What's up with those 1/3 electric charges?
- Are there more particles?

These are all natural questions to ask. And while all this mystery might frighten some people, it's important to take a deep breath. Remember that our strategy is to organize what we know and then look for patterns and holes that we can use to ask the right questions. Asking the right questions will hopefully lead us to a deeper understanding of what's going on.

Decades ago, this table of fundamental particles was incomplete. Several of the quarks and leptons had not yet been discovered. But physicists looked at the patterns in the table and used them to go searching for the missing particles. For example, many years ago scientists knew that there had to be a sixth quark because there was an empty spot in the table. Even though it had never been found, people were so confident it existed that it was included matter-of-factly in many textbooks along with its predicted mass. After twenty years, the top quark was finally found (sort of—its mass was much higher than expected, which is why it took so long to find, and which meant all those textbooks had to be rewritten).

And so physicists have proceeded in this way to fill out and study the patterns in this important table. During the last few decades, we have pieced together some answers and, in some cases, more questions.

24 They prefer the term "big boned."

What Are All These Particles For?

The one thing we *do* know is that there are only three generations of particles. The existence of a fourth generation was ruled out by the discovery of the Higgs boson (see chapter 5 for all your Higgs boson needs). But what does that mean? Is three a basic number in the universe? If you were to finally reveal a single equation that describes everything in the universe, would it have a three in it? Catholics are fond of the number three, but mathematicians and theorists not so much; they like numbers such as zero, one, π, and perhaps *e*. But three? They don't see anything special in it.

What could it mean? We don't have any idea. We literally have no good ideas. There aren't any great competing explanations for the number of particle generations. Quite possibly, it's an emergent phenomenon of some deeper rules about nature, just like the patterns of the periodic table of the elements. Scientists hundreds of years from now might think that we had the clues staring us in the face, that it was *so bloody obvious*, but currently it's a mystery.[25] If you can explain it, find your local particle theorist and knock on her door.

What's the Pattern of the Masses of the Particles?

In the periodic table of the elements, the masses of the atoms and the patterns they formed were a critical clue to figuring out what was going on. From the pattern of the masses, we deduced that each element has a specific number of protons and neutrons in the nucleus (the atomic number, as measured by the positive charge of the nucleus).

25 Future physicists apparently have a condescending British accent.

Unfortunately, there is no apparent pattern to the masses of the fundamental particles. Below are the mass values of each of these particles.

MASS VALUES

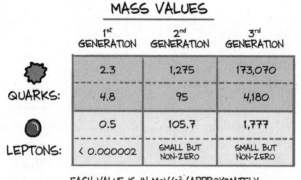

	1st GENERATION	2nd GENERATION	3rd GENERATION
QUARKS:	2.3	1,275	173,070
	4.8	95	4,180
LEPTONS:	0.5	105.7	1,777
	< 0.000002	SMALL BUT NON-ZERO	SMALL BUT NON-ZERO

EACH VALUE IS IN MeV/c^2 (APPROXIMATELY 0.0000000000000000000000000009th OF A CHOCOLATE CHIP)

Other than a general trend that the higher generations are more massive, we haven't been able to figure out any pattern to these values. It might have something to do with the Higgs boson (see chapter 5), but so far there are no clear answers. And take a look at the super-massive top quark. It weighs as much as 175 protons, which is the same as the nucleus of a gold atom.[26] The range of masses spans thirteen orders of magnitude. Why? We have no idea. We are both clueless *and* surrounded by clues.

What's Up with Those 1/3 Electric Charges?

Quarks are unlike leptons in that they feel the strong nuclear force and they have weird fractional electric charges (+2/3 and –1/3). If you mix the up and down quarks in just the right way, you can make protons (two ups and a down quark, with charge = 2/3 + 2/3 – 1/3 = +1) and neutrons (one up and two down quarks, with charge = 2/3 – 1/3 – 1/3 = 0). That's *extremely* important (and lucky) because the charge of the electron just

26 This is supposed to impress you.

happens to be –1. If the quarks had any more (or less) charge, then the charge of protons wouldn't precisely balance the negative charge of the electron and you couldn't form stable neutral atoms. Without those perfect –1/3 and +2/3 charges, we wouldn't be here. There would be no chemistry, no biology, and no life.

This is actually fascinating (or creepy, depending on your level of paranoia) because, according to our current theory, particles can have any charges whatsoever; the theory works just as well with any charge value, and the fact that they balance perfectly is, as far as we know, a huge and lucky coincidence.[27]

Sometimes in science coincidences do happen. The moon and the Sun are vastly different in size, but by cosmic coincidence (this is one of the only times you can scientifically write "cosmic coincidence"), they appear to be nearly the same size in our sky, which allows for dramatic solar eclipses. For ancient astronomers, that must have been quite confusing and suggestive. It likely led many of them down the wrong path trying to understand if the Sun and moon were related in some way. But it is not a perfect coincidence. The sizes of the Sun and the moon in the sky are different by about 1 percent.

Yet, in the case of the fundamental particles, the proton and electron are *exactly* the same charge (but opposite), and we have no idea why. According to our best theory, these numbers could have been anything. This is an exact zero-difference coincidence. What does that mean about the relationship between the electron and the quarks? We don't know, yet it screams for a simpler explanation. If you lost $2,000 on the same day that your neighbor found $2,000, would you chalk that up to coincidence? Probably only after you exhausted many simpler explanations.[28]

It might be that this exact matching of the electric charges is actually another sign that there are deeper components underlying these particles.

27 Some particle theorists argue that of the infinite set of possibilities for the charges, only a few settings give a working theory of physics.

28 Like maybe you should move to a different neighborhood.

Or perhaps these two types of particles are actually two sides of the same coin or built from a common set of super-extra-tiny particle Lego pieces.[29]

Are There More Particles?

In addition to the twelve matter particles (we don't count antimatter particles as unique particles)—the six quarks and six leptons—there are particles that transmit forces. For example, electromagnetic interactions are transmitted via photons. When two electrons repel each other, they are actually exchanging a photon. It's not quite mathematically accurate, but you can think of it as one electron pushing the other away by shooting a photon at it.

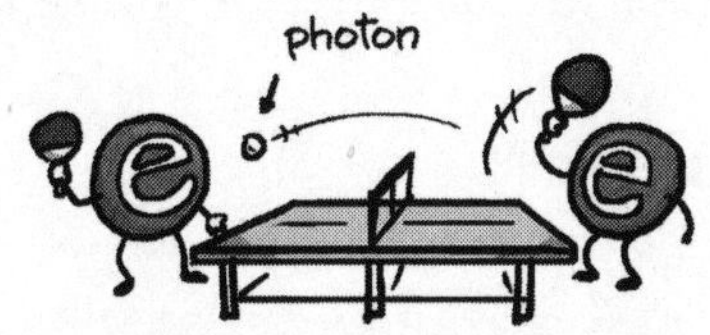

A NOT-THAT-INACCURATE DEPICTION OF PARTICLE FORCE INTERACTION

We know of five force-carrying particles.

FORCE CARRIER PARTICLES

FORCE PARTICLE	FORCE IT TRANSMITS
PHOTON	ELECTROMAGNETIC FORCE
W, Z BOSONS	WEAK FORCE
GLUON	STRONG FORCE
HIGGS BOSON	THE HIGGS FIELD
~~MIDI-CHLORIANS~~	~~THE FORCE~~

29 They still hurt when you step on them.

Combined with our earlier twelve matter particles, this is the full list of particles we have discovered, but we don't know if it's the *complete* list of particles. There is no theoretical limit on the number of particles that could exist. There could be only seventeen particles, or there could be 100, 1,000, or 10,000,000. We know there aren't more generations of quarks and leptons, but there could certainly be other kinds of particles. How many are there? We have no idea.

What Is the Most Basic Element of Matter?

So what are all these particles for? Why are some of them useless if all we need for everyday matter are only the first three (the up quark, the down quark, and the electron)? Well, here are a few possible answers:

- Who knows, but this is it.
- Somebody knows, and this is not it.
- "Useless" is a relative term.

Maybe that's just how the universe is: these particles are the most fundamental objects in the universe, and it just so happens that the universe has a kind of long list of ten to twenty basic parts for no particular reason. Maybe there are other universes out there that have a different list of ten to twenty basic parts, but we may never get to see them.

Or it could be that these particles are not the most fundamental objects in the universe, that they are made of an even simpler set of more basic particles we haven't discovered yet. This means the particles we know are just the result of combining these more fundamental particles. This would explain why there are hints of patterns and coincidences in our current table of particles. This answer is probably the right one, but we have no proof (yet).

Or maybe the heavy particles are "useless" only because they can't be used to make protons, neutrons, and electrons, which are the stable forms of the lightest particles. But the universe is mostly made of these lightest particles only because it is so cold and big. If the universe was smaller and hotter and denser, then we'd have more of the heavy particles and they wouldn't seem so useless (but everything would be very different).

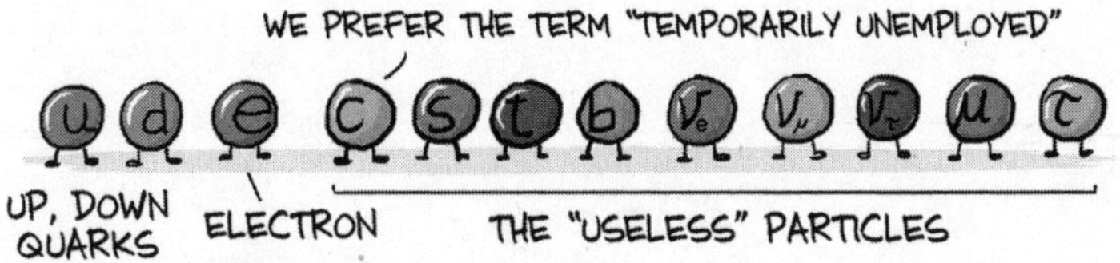
WE PREFER THE TERM "TEMPORARILY UNEMPLOYED"
UP, DOWN QUARKS
ELECTRON
THE "USELESS" PARTICLES

The main takeaway from all of this is that we're still trying to figure out how the 5 percent of the universe we're familiar with works. We've come a long way, but we haven't reached a complete fundamental understanding of why things are the way they are. We have a list of the things we think make up this universe, but we're not 100 percent sure this is the full list.

What's exciting is that we have a solid footing to explore this question. The table of fundamental particles (physicists call it the Standard Model) may have all these unexplained patterns and "useless" particles, but it's based on real observations, and we can use it like a map to discover the true inner workings of the universe. It would be extremely exciting to discover new particles (even if they're not used in everyday matter) because it means we get to expand our map of the universe.

Imagine, for instance, if dark matter is made of particles, ones that we haven't discovered yet. It would open up our understanding of the universe by a whopping 27 percent. In fact, discovering that dark matter is made of only one kind of particle (one that interacts very weakly with our kind of matter) is probably the most boring possible dark matter scenario. Wouldn't it be more exciting if dark matter was made of lots of crazy particles or even a completely different kind of non-particle matter?

The point is that to answer the basic questions of the universe, we have to drill as deeply as possible into the makeup of everyday matter. And along the way, we might dig up particles or phenomena that have no clear role in everyday matter. But we also know that these unexplainable things are part of the universe, so they must hold clues as to why things are the way they are. Answering these questions will fundamentally change how we view ourselves. In other words, we can have our (cosmic) cake and eat it, too.

5.

The Mysteries of Mass

In Which We Take a Light Touch to Some Heavy Questions

You have probably heard it said—by scientists wearing lab coats or, if they are physicists, shorts and T-shirts—that you are mostly empty space. Don't take it personally. What they mean is that the atoms we are all made of have most of their stuff concentrated into a tiny nucleus surrounded by a lot of empty space, making it sound like you should be able to walk through walls.

This is partly true. But the full story is much stranger than that, and it has to do with the many mysteries of "mass." You see, not all the great mysteries of the universe are out there among the stars and galaxies or in strange particles. Some of them are all around you, even inside of you.

We have many descriptions of mass but very little real understanding of what it is and why we have it. We all *feel* mass. As a baby, you develop that sense that some things are harder to push around than others. But as familiar as this feeling is, most physicists would struggle to explain the underlying technical details. As you'll see in this chapter, most of your mass is not made out of the masses of all the particles inside of you. We don't even know why some things have mass and other don't, or why inertia perfectly balances out the force of gravity. Mass is mysterious, and you can't blame it all on that dessert you had last night.

So read on to learn about the many unanswered questions about mass. It would be a massive mistake not to.

The Stuff of Stuff

When you think about things having mass, you probably think about how much *stuff* there is to them. That way of thinking mostly works because you can think of the mass of a typical thing, like a normal, everyday llama, to be the sum of the masses of all the particles inside of it. That is, if you chopped a llama in half,[30] the mass of the llama would be the sum of the masses of the two halves. If you chopped the llama into four pieces, its mass would be the sum of the masses of the four pieces. And so on. If you chop the llama into n pieces, you can measure its mass by adding up the masses of the n pieces. Right?

Wrong! Okay, mostly right. For $n = 2, 4, 8 \ldots$ up to $n = 10^{23}$ or so, it works. But then it doesn't. The reason is going to sound very strange: the total mass of the llama is not just the mass of the stuff inside of it. *It also includes the energy that holds that stuff together.* That's a pretty weird idea; give it a minute to settle in.

30 Llama thought experiment. Don't try this at home.

If you have never heard of this concept before, you are probably hoping that it's just a semantic ploy, that we're using the word "mass" in some technical way to mean something different from the common understanding of mass. The short answer is: no, we mean exactly what you think we mean, but mass is not quite what you thought it was.

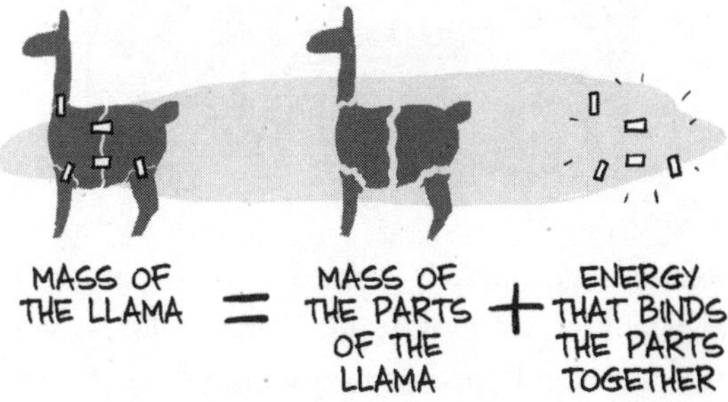

The longer answer requires that we be very clear about what we mean by mass. Mass is the property of objects that makes them resist changes in velocity. Simply put, if you push on something, it will accelerate (change its velocity). But if you push on different things with the same amount of force, you will notice that some accelerate a lot and some accelerate hardly at all. Try this at home by shooting a Nerf gun at things you find around your house, such as tissues and sleeping elephants. Each Nerf bullet applies a nearly equal amount of force, but the effect on the tissue is much greater than on the sleeping elephant.[31] This is what we call mass.

This is also your experience of mass in the everyday world. There's no trickery here. An elephant has more mass than a tissue; that's not *why* it's harder to move; that's what it *means* to have more mass: you get accelerated less by the same force. This is sometimes called "inertial mass" because this quality of resisting acceleration is also known as inertia. We can measure inertial mass fairly easily by applying a known amount of force and measuring the acceleration. (Note that there is a second definition of mass, "gravitational mass," which we will discuss later.)

31 This might depend on what part of the elephant you hit. On second thought, don't try this experiment at home either.

WE UNDERSTAND YOUR CONCERNS,
BUT IT'S FOR SCIENCE!

Now that we have carefully defined what we mean by mass, we can use that definition to measure the llama's mass at any time with government-issue sets of Nerf guns calibrated by NASA engineers. With this in hand, we can turn back to our thought-llama that has been atomized to advance the cause of science.

When you break the bonds that hold the llama's atoms together, you release the energy in those bonds and the total mass of the sliced llama goes down. For $n = 2$ llama pieces, you can't really notice. But if you completely atomize the llama, then it starts to add up. The energy that is stored in the bonds between the bits of llama actually gives the llama more mass. This is not a theoretical conjecture, but an experimental observation.[32]

In the case of a llama, it's not that large an effect. For example, if you broke all the chemical bonds that tie the llama's atoms together, there wouldn't be a big difference between the mass of the llama and the sum of the masses of all of the llama's atoms. And even if you broke up all of the individual atoms into their constituent protons, neutrons, and electrons, there still wouldn't be a big difference in mass (it'd be on the order of 0.005 percent).

With smaller particles, it's a different story. If we were to separate each of the llama's individual protons and neutrons into their constituent

32 Nobody has successfully atomized a llama, but similar experiments have been done. (For the record, we do not support the atomization of llamas. Unless you decide to name your Peruvian punk-rock group The Atomization of Llamas. In that case, we love you.)

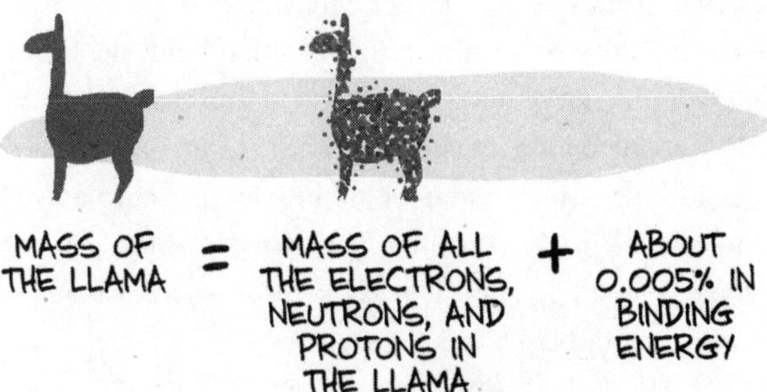

quarks (remember that each proton and each neutron is made of three quarks), we would see a *huge* difference in mass. In fact, *most* of the mass of a proton or neutron comes from the energy that's binding their three quarks together.

In other words, if you were to add the masses of three quarks (measured by hitting each of them with a Nerf gun) and compare that to the mass of those same three quarks bound together in a proton or neutron (measured by hitting the proton or neutron with the Nerf gun), you would see a very big difference in mass. The masses of the individual quarks only account for about 1 percent of the mass of the proton or neutron. The rest is in the energy that's keeping those quarks together.

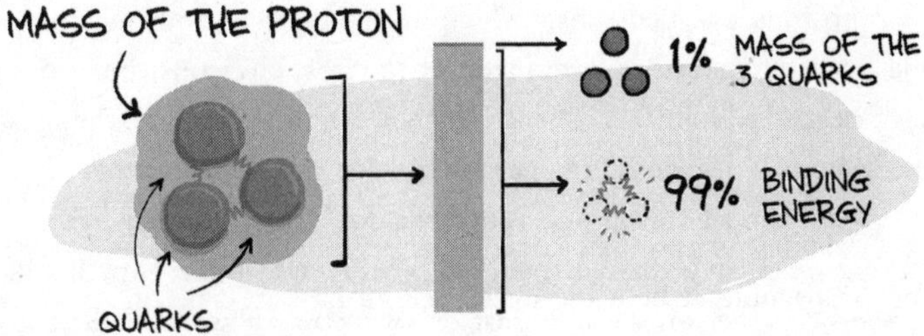

These examples show you what happens when there is energy stored in the bonds between particles: it makes the combined object more massive than the sum of its parts.

To see how strange that is for your intuition, imagine that you took three beans and measured each of their masses. What's the mass of the three beans? It's the sum of the three masses. Easy so far. Now imagine that you put the three beans into a little bag that holds the beans together really tightly with a lot of energy. You would find that all of a sudden the bag would feel much more massive than just the mass of the beans inside of it. It would weigh more, and it would be a lot harder to move from one point to another. What's happening is that most of the mass of the bag doesn't come from adding up the masses of the beans inside but from the energy needed to hold the beans together.

IF YOU SEPARATE THEM THEY'LL RELEASE AN ENORMOUS AMOUNT OF ENERGY!

JACK AND THE BEANSTALK: AN ELABORATE METAPHOR FOR THE PHYSICS OF MASS.

What's crazy is that most of your body is made out of these bags of beans (protons and neutrons), which means most of your mass doesn't come from the "stuff" you're made of (quarks, electrons) but from the energy needed to hold your "stuff" together. In our universe, the mass of something includes the energy needed to keep that stuff together.

And the mind-blowing part is that we don't really know why.

What we mean is that we don't really know why the energy that holds the beans together affects how fast or slow something accelerates in response to a force. If you were to push on your little bag of beans, there's no real reason why you should be able to feel that energy inside. It shouldn't matter to you whether the beans are held together with spit or Super Glue. And yet it does. That's one of the great mysteries of mass. Even though we can measure it, we don't really know what inertia is or why it's tied to both the mass of the particles and the energy that binds the

OUR KNOWLEDGE OF INERTIAL MASS

particles together. You could say our knowledge on this subject amounts to a hill of beans.

Particularly Confusing Particle Masses

If your mind isn't already blown from learning that physics can't really explain something as basic as inertia, get ready for another massive revelation: even the mass that we assign to basic particles like the quark or electrons isn't really "stuff" either. In fact, there is no such thing as "stuff." It doesn't exist in our formulation of physics.

Particles—in our current theory—are actually indivisible points in space. That means that in theory they take up *zero* volume and they are located at exactly one infinitesimal location in three-dimensional space. There's actually no size to them at all.[33] And since you're made of particles, that means you're not mostly empty space, you are *entirely* empty space!

Think for a moment about how much sense that *doesn't* make for the concept of mass. Remember that some particles have tiny almost-zero masses and others have enormous fat masses. For example, here's a question that makes little sense: What is the density of an electron? An electron has nonzero mass and it exists in zero volume, so the density (mass divided by volume) is actually . . . undefined? It makes no sense.

33 There are some definitions of particle size that incorporate the virtual particles that surround them, but we take a stricter approach.

Or take two particles that are identical in every way other than mass, such as the top quark and the up quark. The top quark is like the up quark's superfat cousin; it has the same electric charge, the same spin, and the same interactions. They are both supposed to be fundamental point particles, but the top quark is *75,000 times* more massive. And yet they take up the same amount of space (none) and act almost the same way. So how does one of them have more mass than the other without having any more "stuff" to it?

The reason this seems to make no sense is because particles are unlike anything else you have experienced in your day-to-day world. It's totally natural that when we try to understand something new we use models that are based on things we know.[34] What else could we do? It's like explaining to a three-year-old what a tiger is. You might say it is "just like a big kitty cat," but that only works until the three-year-old is at the zoo one day and tries to stick her hand in the cage to pet the tiger and your spouse yells at you for being a bad parent who uses theoretically incomplete analogies. These mental models are useful, but you always have to keep in mind their limitations.

We like to think of particles as tiny little balls of stuff. That works for lots of thought experiments even though particles aren't little balls. Not even a little bit. According to quantum mechanics, they are superbizarre little fluctuations in fields that permeate the entire universe. That means they obey rules that make very little sense in the tiny-little-ball model. For example, they can be on one side of an impenetrable barrier one moment and then appear on the other side the next—without passing through the

34 Describing the unknown in terms of the known is the core task of physics. That and making you sound smart at cocktail parties.

barrier.[35] Quantum particles can do things that seem to make no sense if you think of them in terms of things you know because they are unlike anything you have ever experienced.

The models in our head can be useful for giving us intuition or helping us visualize, but it's important to remember that they are just models and they can break down. That's what happens in your brain when you think of the masses of point particles.

WHY?? WHY AREN'T I GOOD ENOUGH?

THE PARTICLE MODEL BREAKS DOWN

Take the other extreme: How does it make sense for a particle to have zero mass? For example, the photon has exactly zero mass. If it has no mass, then it's a particle of *what*? If you demand that mass is equal to stuff, then you have to conclude that a massless particle literally has nothing to it.

Instead of thinking about a particle's mass as how much stuff is crammed into a supertiny little ball, just think of it as a *label* that we apply to an infinitesimal quantum object.

You maybe didn't realize it, but you already think this way when it comes to a particle's electric charge. We all know that electrons have negative electric charge, but when you think about that, do you ever wonder to yourself: *Where* inside the electron is the charge? What is the stuff that gives it the charge, and is there room in the electron for that amount of it? Those questions seem silly because we think of charge as something a particle just has. It's a label, and it can have lots of values: 0, –1, 2/3, etc. Try to think of mass the same way, and it will make a little bit more sense.

35 Quantum tunneling. A phenomenon so well established it is used routinely in some supermicroscopes. It really happens.

WHOA. YOU NEED TO LOSE WEIGHT.
1 kg
20 kg

But if electric charge means a particle can feel electrical forces (like getting repelled by other electrons), what does mass mean for a particle? Mass is the thing that gives a particle inertia (resistance to motion). But what we still don't understand is: Why do things have inertia at all? Where does it come from? What does it mean? Who will help us in our hour of need? The answer is: the Higgs boson.

The Higgs Boson

In 2012 particle physicists announced the discovery of the Higgs boson to great international fanfare. Almost nobody understood what the Higgs boson was, but lots of people got very excited. The *New York Times* wrote that it "represents the very best of what the process of science can offer to modern civilization." That's right, the Higgs boson is apparently better than computers, flushing toilets, and reality TV.[36]

So what is the Higgs boson? Here's a quiz to test your knowledge. Take it now and then again after you read this chapter. We hope that at the very least your score will not decrease.

36 We concede that the Higgs may be more important than at least *one* of these.

THE HIGGS QUIZ
i.e., "The Hizz"

1. Before it was reused as the name of a particle, "Higgs boson" was famous as:
   a. A beloved children's television clown
   b. The code name of the CIA's most dangerous spy
   c. Luke Skywalker's childhood friend in *Star Wars*
   d. Your friend's Dungeons & Dragons character

2. True or false: if consumed directly, the Higgs boson is more addictive than Flamin' Hot Cheetos.
3. True or false: the Higgs boson is a particle predicted by two theorists named Higgs and Boson.

Check your answers in the footnotes to see how much you know.[37]

In all seriousness, finding the Higgs boson *was* a triumph of science. It was a demonstration that looking for patterns is a good guide to understanding the universe.

The idea that the Higgs boson might exist came out of studying the patterns of the particles that transmit forces—the photon, the W boson, and the Z boson—and asking questions about their mass. Physicists asked: Why is one of them massless (the photon) and the others (W and Z) very massive? It didn't make sense in this particular case for this strange

I'VE ALWAYS BEEN LIGHT.

W
Z
Photon

37 If you actually answered any of these questions, it's probably good that you're reading this chapter.

label that we call mass to be zero for one force particle and yet be nonzero for the others.

Peter Higgs and several other particle physicists stared at this for a while until they found the solution: just make stuff up. Literally. They posited that if you add one more particle (the Higgs boson) and its field (the Higgs field) to the equations then mass as a particle label—and why some particles have it more than others—start to make sense.

Roughly, the theory goes like this: imagine a field that permeates the entire universe. This field does something no other field does: rather than attracting or repelling anything, it makes it hard for particles to get going or slow down. The effect of this field is *identical to the effect of having inertial mass*.

The more the field interacts with a particle, the more it seems that the particle has inertia—or has mass. It goes one step further and suggests that the inertia generated by a particle interacting with this field *is* the particle's mass. That's what it means to have mass. Some particles feel this field very strongly, meaning they take a lot of force to speed up or slow down; these particles have a lot of mass. Other particles hardly feel this field so they take very little force to speed up or slow down; these particles have almost no mass. According to the Higgs theory, that's what mass is.

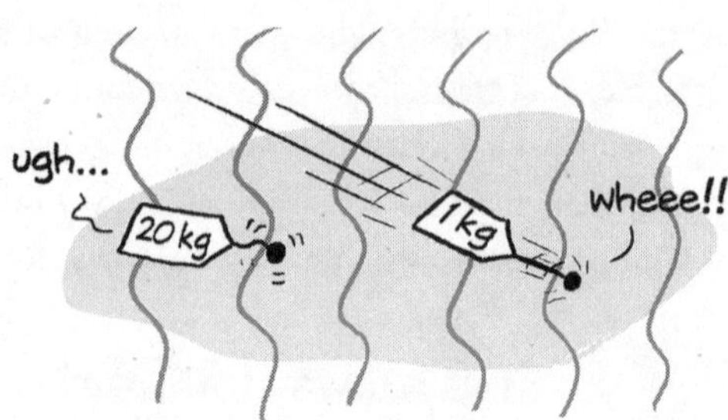

Take a moment to contemplate that. It's both a paradigm-changing insight and a *totally trivial statement* at the same time.

It's paradigm-changing because it gives you a different idea of what mass is. That's kind of a big deal.

But it's also trivial because, once you accept that mass is a mysterious quantum label for a particle rather than the amount of stuff inside of it, learning that the size of that mass label comes from a mysterious universe-spanning field doesn't help you understand what mass is.

In fact, it does nothing to address the most important question: Why do the matter particles have different masses? The Higgs theory says that the reason is that they feel the Higgs field differently. So all the theory does is turn one question into a different question: Why do all the matter particles feel the Higgs field differently?

I BLAME THE HIGGS FIELD.
W
P

According to the theory, there is no rhyme or reason to the masses of the matter particles. It's as if they were randomly selected and they could just as well have had totally different values. Nothing in our theory would break if you changed the masses. The same laws of physics we have now would work just as well. Of course, making some of the particles more or less massive would have big effects on other things, such as the protons, neutrons, and electrons that we count on for making our overpriced seasonal lattes (and chemistry and biology more generally). But according to the current theory, the masses of the matter particles are arbitrary parameters, free to be set to any value.

The Higgs theory *does* explain why the force particles (photon, W, and Z) have the masses they do, but it doesn't generally explain why the matter particles have different masses (why some interact with the Higgs field a lot and others don't). There is probably a pattern to the masses, but it's one that has so far escaped us. Our level of sophistication is just like Ook

MAMA ALWAYS SAID:
MASS IS AS MASS DOES.
H

and Groog's, who explained things by listing them. In the same way, our best theory of the universe only lists the masses of the matter particles as arbitrary numbers.

Perhaps some future scientist will look at our list and roll her eyes at our ignorance as she writes down a simpler theory in which the values of these masses are not arbitrary parameters but rather the result of some deeper, more beautiful description of nature. We still have no idea.

Gravitational Mass

This brings us to the final piece of the puzzle.

When we thought earlier about how to measure the mass of something, you might have had a different idea than our precision Nerf-gun approach: just use a scale! A scale measures the weight of an object, which means the gravitational pull of the Earth on it. That's very closely related to mass, because the more mass something has, the stronger the Earth pulls on it. The force of the Earth on an elephant is greater than the force of the Earth on a tissue.

In the case of a particle, you can also think about gravitational mass as a gravitational *charge*. When two particles have electric charges, they feel electrical forces on each other, and the electrical force is proportional to the charges. In the same way, when two particles have mass, they feel a gravitational attraction proportional to their masses.

1kg

20kg

GRAVITY ONLY ATTRACTS

Oddly enough, you can't have negative mass, so there's never gravitational repulsion, only attraction.[38] Gravity is different from other forces that way, which we will explore in more detail in the next chapter.

38 *Almost* never. Dark energy and inflation might be due to gravitational repulsion.

Are the Two Kinds of Mass the Same?

Is gravitational mass the same as the inertial mass we were talking about a few pages back? Yes . . . and no.

No, because this mass that we call "gravitational mass" seems to determine the force of gravity on an object, and we measure it using a different technique (a scale) than we do with inertial mass.[39]

MASS DEMONSTRATIONS

And, yes, because we can measure the mass both ways, and so far we have never observed *one iota of difference* between the gravitational and inertial masses of an object.

Think about how weird that is. There's no real intuitive reason why the two should be the same. One of them (inertial mass) is how resistant something is to being moved, and the other (gravitational mass) is how much it *wants* to be moved by gravity.

You can do a simple experiment to confirm this. Drop two objects with different masses (like a cat and a llama) inside of a vacuum (so there is no air resistance) and you will see that they fall at the same speed. Why does that happen? If the gravitational mass of the llama is larger, then it gets pulled by a larger force from the Earth; but since the llama also has a larger inertial mass, it takes a larger force to get it moving. The two effects perfectly cancel out each other, and the cat and llama fall at the same speed.

39 This is the Newtonian view of gravitational forces. Later, we'll get general relativity's version, in which there are no gravitational forces and it makes more sense to think of mass as distorting space.

In our current formulation of physics, we don't know why that is. We just assume it is the same. And this assumed equivalence is at the heart of Einstein's general theory of relativity, which looks at gravity in a very different way. Rather than thinking of it as a force that acts on an arbitrary charge attached to particles and the energies that bind them, it describes gravity as the bending or distorting of space around both mass and energy. So in Einstein's theory the connection is much more natural, but it still doesn't tell us *why* it's there. Are there two arbitrary parameters (inertial and gravitational mass), or are they connected? Could the two have been different without breaking the laws of physics?

Other than relativity, our particle physics theories treat gravitational and inertial masses as different concepts, but experimentally we see them as the same thing. That's a very strong suggestion that they are deeply connected.

Heavy Questions

To recap, here are the ways in which mass is weird:

It's weird because the mass of something is not just the mass of the stuff inside of it. Mass also includes the energy that binds the stuff together. And we don't know why that is.

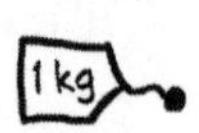

It's weird because mass is actually like a label or a charge (it's not really "stuff"), and we don't know why some particles have it (or feel the Higgs field) and others don't.

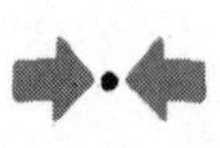

And it's weird because mass is exactly the same whether you measure it via inertia or gravity. And we don't know why that is either!

What's interesting is that, for all the mysteries of mass, it has actually helped us make progress in understanding the rest of the universe. Remember that it was the rotation of galaxies and the problem of missing mass that gave us the clue that there was an invisible new kind of mass in the universe: dark matter. In fact, just about the only thing we know about dark matter is that it has mass: gravitational mass, to be precise.

It's amazing to think that something so fundamental to our existence can still be a mystery. What are we paying all those physicists for if not to help us sleep better at night knowing these kinds of things are taken care of? But, no, the more you probe into it and ask questions, the more you realize there are still things about mass that are puzzling.

What's clear (and exciting) is that mass is a fundamental property of how the universe works and that it clearly connects a lot of its moving pieces (energy, inertia, and gravity, for instance). Finding out exactly what those connections are would bring us another step closer to understanding this big and wonderful universe we live in. And that would be (okay, last one) *massively* cool.

6.

Why Is Gravity So Different from the Other Forces?

It's a Big Question of Little Gravity

You *know* what gravity is. It controls the motion of stars, creates black holes, and drops apples on the heads of famous but clueless physicists.

But do you really *understand* gravity?

You see it working around you, but when we compare the way that it works to the patterns set by the other basic forces, we notice immediately that it doesn't quite fit. It is weirdly weak, it nearly always attracts rather than repels, and it doesn't play nice with a quantum view of the world.

And that refusal to fit in is very mysterious and frustrating because finding patterns is how we understand the universe. Look around you and you might be overwhelmed by the variety and complexity of our beautiful universe, but find the patterns and you can begin to make sense of it. For example, think about how much you can understand about a person if you examine the patterns in their Internet browsing history. Then again, maybe that's a part of the universe you don't want to understand.

NSA
THE "BIG BANG"? "BLACK HOLES"? THAT IS SOME RACY BROWSER HISTORY.

But the desire to fit things into patterns in order to understand them is the reason that physicists salivate over the idea of unifying all of physics into a single theory.[40] And gravity's refusal to fit into the pattern of all the other forces is a big obstacle to achieving that. In this chapter, we'll explore why exactly gravity is so peculiar and why it's pulling down more than just your average papaya or llama onto the Earth. There are deep mysteries about gravity, so let's get started and fall right into them. We might even gravitate toward some answers.

These puns are getting heavy.

Gravity's Weakness

Everyone at some point wonders, *Why am I here on this Earth?* We have the answer: gravity. Without gravity, we would all float off into space and the universe would be a dark, giant, amorphous cloud of dust and gas. There wouldn't be any planets, stars, silly tropical fruits, galaxies, or good-looking people who buy humor-tinged books about physics. Gravity is huge. But it's also really *weak*.

How weak is gravity? Well, roughly speaking, gravity is about 10^{36} times weaker than the other three fundamental forces. That's a fraction of 1/1,000,000,000,000,000,000,000,000,000,000,000,000.

How can we understand a number like that? Let's borrow a strategy from how we learned fractions in first grade. If you had a papaya and cut it into four pieces, each piece would be a quarter of a papaya. Easy. If you had a papaya and cut it into 10^{36} pieces, each piece would be . . . less than a single papaya molecule.[41] In fact, you would need to cut up about *two*

40 Let's be honest, physicists salivate easily.

41 Papaya molecules are called papayons, and they are tiny and sweet.

million papayas to make a fraction of $1/10^{36}$ equal approximately one papaya molecule.

A good way to see the weakness of gravity is to do a little experiment that pits it against other forces. You don't need a particle accelerator in your basement for this. Just take a standard kitchen magnet and use it to lift a small metal nail. In your experiment, that nail is being pulled down by the gravitational force of an entire planet (the Earth), and yet the magnetic force from a tiny little magnet is enough to keep the nail from falling. A tiny magnet overpowers a whole planet because magnetism is so much more powerful than gravity.

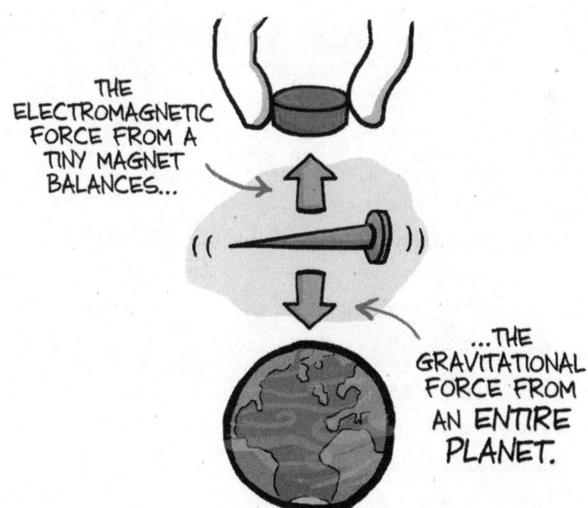

At this point, you might be wondering: If the force of gravity is thirty-six *orders of magnitude* weaker than all the other forces, how can it be so consequential in our universe? Won't it be blown away by the more powerful forces around it, like a sneeze in a tornado?[42] How is it keeping all the planets and stars together and how is it keeping everyone from flying around like Superman? If the other forces are so strong, wouldn't they overwhelm over gravity and totally wash out its effect on the universe?

The answer is that gravity is very important at huge scales and when dealing with enormous masses.[43] The weak and strong forces are short-range forces, so they are mostly felt only at the subatomic level. And the reason electromagnetic forces don't play a large role in the movements of stars and galaxies, even though those forces are huge compared to gravity, has to do with an interesting fact about gravity: it mostly works only one way.

Gravity works to only pull things together, not push things apart.[44] The reason is simple: gravity's force is proportional to the mass of the

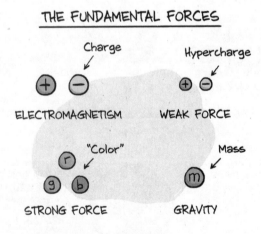

THE FUNDAMENTAL FORCES

Charge
ELECTROMAGNETISM
Hypercharge
WEAK FORCE
"Color"
STRONG FORCE
Mass
GRAVITY

42 Also true for a fart in a hurricane.

43 Gravity likes big masses and it cannot lie, the other forces can't deny.

44 Almost always true: see chapter 14 for a discussion of repulsive gravity during the Big Bang.

objects involved, and there is only one kind of mass you can have—positive. In contrast, electromagnetic forces have two kinds of electric charge (positive and negative), and the weak and strong forces have properties just like electric charge called hypercharge and color, which can also have multiple values.[45]

Gravity is sort of the same way, but not quite. You can think of mass as the "gravity charge" of a particle that determines how much gravity it feels. But there is no "negative" mass. Gravity doesn't repel particles with mass.

This is important because it means that gravity can't be canceled out. This is what happens to the electromagnetic force at large scales. If the Sun was made up of mostly positive electric charges, and the Earth was made up of mostly negative electric charges, the attraction would be *enormous* and our planet would have been sucked into the Sun a long time ago.

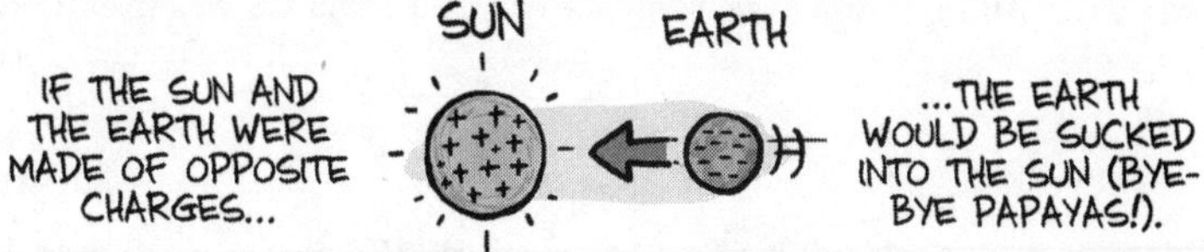

SUN
EARTH
IF THE SUN AND THE EARTH WERE MADE OF OPPOSITE CHARGES...
...THE EARTH WOULD BE SUCKED INTO THE SUN (BYE-BYE PAPAYAS!).

But because the Earth is made up of almost equal amounts of positive and negative charges, and the Sun is also made up of almost equal amounts of positive and negative charges, they mostly ignore each other electromagnetically. Every positive and negative particle on the Earth is both attracted and repelled by the positive and negative charges in the Sun (and vice versa), so all the electromagnetic forces cancel out.

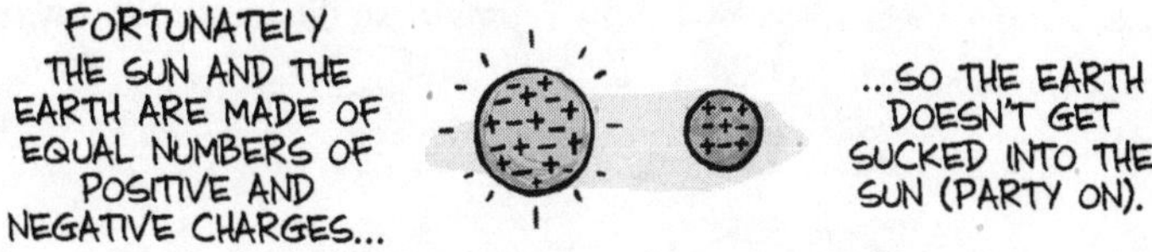

45 The strong force has more than two types of charge. It has three! They are called "colors" and named "red," "blue," and "green." To cancel a red charge, you can either add a blue and a green particle to get a neutral or "white" object, or you can find an antiparticle that is colored antired.

This is no accident. The electromagnetic force is so powerful that it will suck charges back and forth until any residual imbalance disappears. It was very early in the lifetime of the universe (when it was 400,000 years young, in the pre-papaya period) that virtually all matter settled into neutral atoms and electromagnetic forces found balance.

Since there is no net electromagnetic force between the Earth and the Sun, and since the weak and strong forces don't work on this distance scale, the only force left is gravity. This is why gravity dominates at the scale of planets and galaxies: because all the other forces are in balance. Despite being so attractive, gravity is like the last one left at the party holding a papaya when everyone else has found someone to go home with. And because gravity only attracts, it never cancels itself out.

So there are two curious and as-yet-unexplained properties of gravity: first, it is really, really weak compared to the other fundamental forces. Imagine that everyone else brought a light saber to a fight and gravity brought only a toothpick. The other curious property about gravity is that it only attracts. All the other forces attract or repel depending on the charges of the particles involved. Why is gravity so different in this way? We have no idea.

The Quantum Conundrum

Gravity almost, but not quite, fits in the pattern set by the other three fundamental forces. We can think of it as a force like all the others, and we can think of mass like we think of the other charges. But gravity is much weaker and only works in one direction. This apparent inconsistency in the forces means either the pattern we have is not valid or we are missing something big.

It turns out that gravity is weird in other, more profound ways, too. We have a way of understanding all the matter particles and three of the four fundamental forces in a mathematical framework called quantum mechanics. In quantum mechanics, everything is described as a particle, even these three forces. When an electron pushes on another electron, it doesn't use the Force or some form of invisible telekinesis to cause the other electron to move. Physicists think of that interaction as one electron tossing another particle at the other electron to transfer some of its momentum. In the case of electrons, these force-carrying particles are called photons. In the case of the weak force, particles exchange W and Z bosons. Particles that feel the strong force exchange gluons.[46]

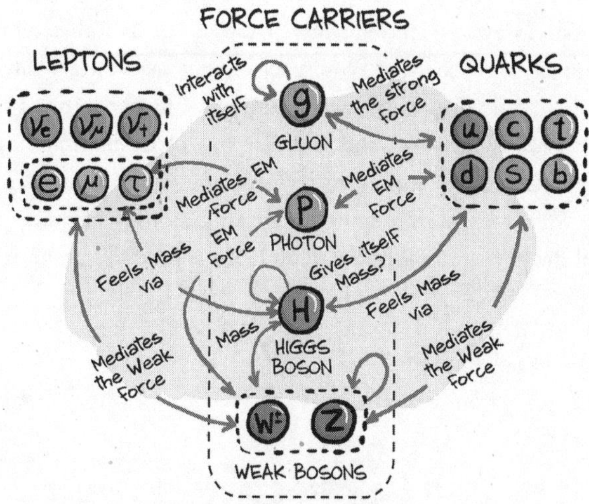

PARTICLE INTERACTIONS: MORE COMPLICATED THAN A MEXICAN TELENOVELA

46 You probably don't believe us because we made up "papayons," but gluons are real!

This quantum mechanical framework, the Standard Model of particle physics from chapter 4, has been incredibly successful at describing most of the natural world (by "most" we mean a whopping 5 percent of the universe, remember?). Looking at the world in terms of quantum particles can explain many things we have seen in experiments, and it has allowed us to predict things we had never seen before, like other matter particles or the Higgs boson. It even explains why the weak force has such a short range: its force particles have a lot of mass, which limits how far they can travel. But there is a big problem with the Standard Model: the same approach doesn't quite work to describe gravity.

Graviton: Elementary Particle or Comic Book Supervillain?

Quantum mechanics fails to describe gravity for two reasons. First, fitting gravity into the Standard Model requires a particle that transmits the force of gravity. Physicists have creatively called this hypothetical particle the "graviton." If it exists, it would mean that, as you are sitting (or standing) there being pulled down by gravity, all the particles in your body are constantly throwing and receiving tiny little quantum balls with all the other particles of the Earth beneath you. And as the Earth goes around the Sun, there's a constant stream of gravitons being exchanged between all the particles on the Earth and all the particles in the Sun. The

YOU'RE NOT LAZY, YOU'RE JUST BEING HELD DOWN BY EVERY SINGLE PARTICLE IN THE ENTIRE PLANET.

problem is, nobody has ever seen a graviton, so this theory could be completely wrong.

The other reason physicists have trouble incorporating gravity into quantum mechanics is that we already *have* a great theory of gravity, one that Einstein came up with in 1915. It's called general relativity, and it works pretty well on its own. It looks at gravity in a totally different way: rather than thinking of gravity as a force between two objects, Einstein looked at gravity as a distortion of space itself. What does that mean? Einstein realized that gravity becomes simple if you stop thinking about space as an abstract concept, the invisible backdrop for all matter, and instead think of it like a dynamic fluid or flexible sheet. The presence of matter (or energy) bends space around it, changing the path of objects. In Einstein's picture, there is no force of gravity, only a distortion of space.

PLAUSIBLE THEORIES ABOUT GRAVITY:
IT'S A DISTORTION OF SPACE-TIME.
IT'S MEDIATED BY QUANTUM GRAVITONS.
IT'S THE WARM EMBRACE OF THE GIANT SPAGHETTI MONSTER.

According to general relativity, the reason the Earth goes around the Sun rather than flying off into space is not because there is a force that pulls it around in an orbit. It goes around the Sun because the space around the Sun is distorted in such a way that what feels like a straight line to the Earth is actually a circle (or an ellipse). In this scenario, gravitational mass is not a charge that some particles have and others don't; rather it is a measure of how much an object is capable of distorting the space around it. As bizarre as this theory may sound, it's been very successful at describing local gravity, cosmic gravity, and many other strange things we see out in space. It explains why light bends around objects and why your GPS works, and it predicted black holes.

The problem is that general relativity works very well, so we think it's

probably a correct description of nature, but we haven't been able to merge it with that other fundamental theory, quantum mechanics, which *also* seems like a correct description of nature.

Part of the problem is that they look at the world so differently. Quantum mechanics views space as a flat backdrop, but general relativity tells us that space is part of a dynamic, flexible thing: space-time. So is gravity a distortion of space, or is it little quantum balls flying around between particles? Everything else in our universe is quantum mechanical, so it would make sense if gravity followed the same rules, but so far there is no evidence to convince us that gravitons exist.

GRAVITY CAN BE A DOWNER.

Even more problematic is that we can't even predict what a merged theory of quantum gravity would *look* like. Physicists have often been able to predict particles that were later discovered experimentally (like the top quark or the Higgs boson), but so far all the theories we have that try to merge gravity and quantum mechanics fail; they keep giving nonsense results, like "infinity." Theorists are a smart bunch (in theory), and they have some good ideas that might one day lead to a merged theory—such as string theory or loop quantum gravity—but it's fair to say that to date the progress has been slow. See chapter 16 for more discussion of theories that unify all knowledge.

Black Hole Colliders

To summarize, gravity appears to be so different from the rest of its force siblings that everyone speculates it was either adopted or the result of some funny business by Mrs. Universe. Gravity is much smaller than the other forces, it only works one way (attracts, not repels), it doesn't seem to fit into the same theoretical structure as the other forces, and we have no idea why. These are some of the biggest mysteries in the universe. What are we doing to answer these conundrums?

One approach to understanding how the world works is to test it with experiments and then come up with clever ideas that explain what we have seen. Ideally, we would like to test general relativity (classical gravity) and quantum mechanics at the same time to see which is correct (if either) and which breaks down. For example, observing two masses exchanging a graviton would demonstrate conclusively that gravity is a quantum phenomenon.

That would be great, but think about how difficult that experiment would be. Remember that gravity is really weak. Not even the gravity of the entire Earth is enough to overpower the electromagnetic force of a tiny magnet. If you were to bring together two particles, the gravitational force between them would be almost zero, and it would be blown away by the more powerful electromagnetic, weak, and strong forces.

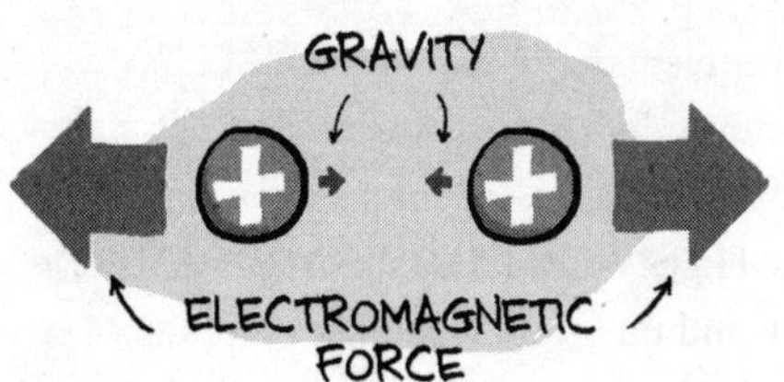

In order to observe gravitons, we need a lot of mass. We need an experimental situation in which we are colliding enormous cosmic-size masses that are balanced in all the other forces. No, we are not thinking of colliding a million kilograms of papayas.[47] Stretch your imagination to the breaking point and try to picture in your mind the incredible concept of a *black hole collider.*

Two cosmically massive objects slamming into each other—that's what you'd need to probe gravity at the quantum level. Clearly, this is not something that one can build or operate (reasonable budget estimates would make the Death Star look cheap). However, we are lucky that the universe is a very big place full of very weird stuff. If you look around long enough, you can find almost anything you are looking for, including colliding black holes.

These events don't happen on schedule and are not repeatable, but every so often, black holes get close enough to each other that they try to suck each other in. This is exactly what scientists are looking for. There are places in the cosmos where black holes are engaged in a death spiral and the collision might be generating gravitons shooting out in every direction. All we need to do is see them! It turns out that is not so easy. Even the gravitons produced by a black hole collider will be very hard to spot. The weakness of gravity means that even if a graviton passed through you, you would hardly feel it. Remember neutrinos, the ghostlike particles that can pass through light-years of lead? Gravitons make neutrinos seem like social butterflies that like to talk to everyone at the party. In fact, one

BLACK HOLE DEATH MATCH

47 Okay, now we are thinking about it.

calculation suggests that a detector the size of Jupiter would see one graviton every ten years even if it was near an intense graviton source.

Let's Get Realistic and Visit a Black Hole

If seeing an individual graviton is impossible, how can we possibly understand whether gravity is a quantum theory or not? Another way to do it is to find a physical situation where the two theories disagree in their predictions. For example, a slightly less unrealistic scenario is to explore the *inside* of a black hole.

General relativity tells us that at the heart of a black hole there exists a singularity, a point where matter is so dense that the gravitational field becomes infinite. This would be a (literally) mind-bending experience because space-time would distort you beyond any intuitive understanding. General relativity has no problem with such a thing existing, but quantum mechanics disagrees. According to the principles of quantum mechanics, it's impossible to isolate anything exactly to a single point (like a singularity) because there is always some uncertainty. So one of the two theories has to break down in this situation. If we knew what was actually happening inside a black hole, we would have some very important clues about how quantum mechanics and gravity play together. Unfortunately, the prospects for visiting a black hole, surviving it, doing the experiments, escaping the inescapable gravitational field, and returning to Earth with the results seem daunting at this point.

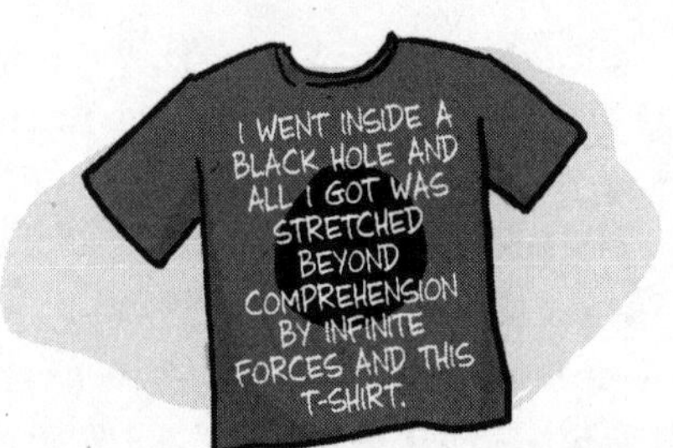
I WENT INSIDE A
BLACK HOLE AND
ALL I GOT WAS
STRETCHED
BEYOND
COMPREHENSION
BY INFINITE
FORCES AND THIS
T-SHIRT.

BLACK HOLES MAKE THE WORST VACATION DESTINATIONS

So Much for That

But even if we can't use them to discover gravitons, we can still learn things from a black hole death spiral because they can produce *gravitational waves.*

Gravitational waves are the ripples in space caused by accelerating masses. It's similar to what happens when you put your hand into a bathtub full of water and move it back and forth. Your hand will send ripples in the water down to the other end of the tub. The same thing happens when massive objects move in space. The moving mass bends space itself, creating a disturbance that can propagate like a wave.

The cool thing is that when a gravitational wave passes by, everything along its path is stretched and distorted. A circle will momentarily become an ellipse, and a square will become a rectangle. Sounds cool, right? Before you stop reading to see if this book is changing shape, you may want to know that gravitational waves distort space only by about a factor of 10^{-20}. That means if you had a stick that was 10^{20} millimeters long (ten light years), a gravitational wave would shorten it by one millimeter. That's a difficult effect to measure.

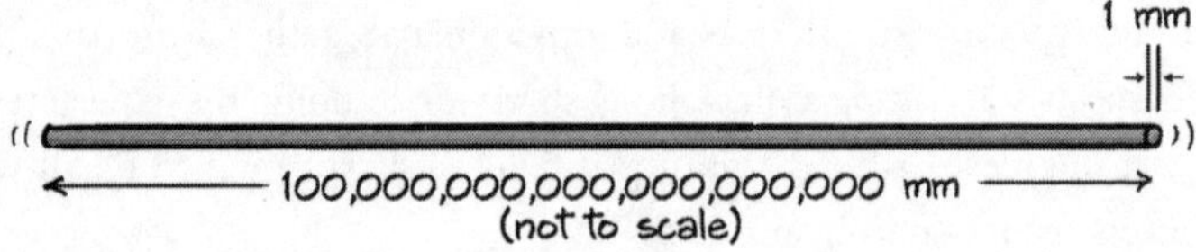

But scientists can be clever and patient. They constructed an experiment called LIGO (Laser Interferometer Gravitational-Wave Observatory). It has two four-kilometer-long tunnels at right angles to each other, and uses a laser to measure the changes in the distance between the ends of the tunnels. When a gravitational wave comes through, it stretches space in one direction and squeezes space in the other direction. By measuring the interference of the lasers as they bounce between the different ends, physicists can measure very precisely whether the space in between has stretched or compressed.

THE LIGO EXPERIMENT
THE LEGO EXPERIMENT
$!?
LEGO
MEASURES GRAVITATIONAL WAVES.
MEASURES HOW MUCH PARENTS ARE WILLING TO PAY FOR SMALL PIECES OF PLASTIC.

In 2016, after $620 million and decades of watching, scientists spotted their first gravitational wave. This beautifully confirmed Einstein's picture that gravity bends space itself. Unfortunately, it doesn't give us any insight into a quantum picture of how gravity works because gravitational waves are not the same as gravitons. It's like proving that light exists, but not that it is made of photons. Nevertheless, it was a "massive" discovery and should be treated with great "gravity" (sorry).

Maybe Gravity Is Special

So what are some of the explanations for the mysteries of gravity? Why is it so weak and why does it not fit into the pattern and theories of the other forces?

It may be that gravity is special. There is no rule that gravity *has* to be like the other forces or that there has to be one theory to rule them all. We always need to keep in the back of our minds the larger perspective that we are still in the dark about most of the basic truths about the universe. In many cases, assumptions we have made turned out to be false or only true under certain special conditions. It may be that gravity is totally different from anything we have seen before. Or not. Remember that our goal is to understand the universe, and we should avoid making too many assumptions as to what that looks like.

If it turns out that gravity is special and that it *is* different from the other fundamental forces, that would also be a clue about the bigger picture. It might mean that gravity is something deeper that is ingrained in the fabric of the cosmos. Sometimes we learn more from the exceptions than the rules. And there is no shortage of exciting ideas to explain these mysteries.

One mind-blowing explanation for the weakness of gravity is the idea of *extra dimensions*. Not alternate dimensions like the kind you see in comic books, but more dimensions in space than the ones you currently think you live in. Some physicists are proposing that gravity is weak because it gets diluted into these other dimensions that form little loops that we can't see. If you take into account these extra dimensions, then gravity is just as strong as the other forces. We'll talk about this idea more in chapter 9.

Although we mentioned some of the difficulties in trying to merge quantum mechanics and general relativity, and detecting a graviton, it doesn't mean that physicists have given up on the idea of finding a unified theory that can explain all the forces we know about. How close are we to having a single simple equation that predicts everything? We'll explore that in chapter 16.

What It Could Mean

Understanding the mysteries of gravity would have a huge impact on our understanding of the world around us. Remember that gravity is basically the only force that works on grand scales, which means it's one of the main forces that's determining the shape and eventual fate of the universe.

The fact that gravity bends and distorts space and time could also lead to some very exciting possibilities. Right now, it's quite likely that we will never visit another star system besides our own. The distances are just too far. But if we can understand the mysteries of gravity, it may lead us to understand more about how space can be bent and controlled or how wormholes can be created or manipulated. If that happens, then our wildest dreams of traveling across the universe by folding space-time could become a reality. And gravity may hold the key to that.

Who said gravitational forces always keep your feet on the ground?

PHYSICISTS MAKE HARSH MOVIE CRITICS.

7.
What Is Space?

And Why Does It Take Up So Much Room?

The first several chapters of this book have been about the mysteries of *stuff*: What are the smallest bits and how do they work together to make the universe? But even as we grasp for answers to questions about the tangible things surrounding us, there is a great mystery hanging out in the background. That mystery is the background itself: space.

What is space, anyway?

Ask a group of physicists and philosophers to define "space" and you will likely be stuck in a long discussion that involves deep-sounding but meaningless word combinations such as "the very fabric of space-time itself is a physical manifestation of quantum entropy concepts woven together by the universal nature of location." On second thought, maybe you should avoid starting deep conversations between philosophers and physicists.

Is space just an infinite emptiness that underlies everything? Or is it the emptiness *between* things? What if space is neither of these but is a physical thing that can slosh around, like a bathtub full of water?

It turns out that the nature of space itself is one of the biggest and strangest mysteries in the universe. So get ready, because things are about to get . . . spacey.

Space, It's a Thing

Like many deep questions, the question of what space is sounds like a simple one at first. But if you challenge your intuition and reexamine the question, you discover that a clear answer is hard to find.

Most people imagine that space is just the emptiness in which things happen, like a big empty warehouse or a theater stage on which the events of the universe play out. In this view, space is literally the *lack of stuff*. It is a void that sits there waiting to be filled, as in "I saved space for dessert" or "I found a great parking space."

If you follow this notion, then space is something that can exist by itself without any matter to fill it. For example, if you imagine that the universe has a finite amount of matter in it, you could imagine traveling so far that you reach a point beyond which there is no more stuff and all the matter in the universe is behind you.[48] You would be facing pure empty space, and beyond that, space might extend out to infinity. In this view, space is the emptiness that stretches out forever.

Could Such a Thing Exist?

That picture of space is reasonable and seems to fit with our experience. But one lesson of history is that anytime we think something is obviously true (e.g., the Earth is flat, or eating a lot of Girl Scout cookies is good for you), we should be skeptical and take a step back to examine it carefully. More than that, we should consider radically different explanations that also describe the same experience. Maybe there are theories we haven't thought of. Or maybe there are related theories where our experience of the universe is just one weird example. Sometimes the hard part is identifying our assumptions, especially when they seem natural and straightforward.

In this case, there are other reasonable-sounding ideas for what space could be. What if space can't exist without matter—what if it's nothing more than the *relationship* between matter? In this view, you can't have pure "empty space" because the idea of any space at all beyond the last piece of matter doesn't make any sense. For example, you can't measure the distance between two particles if you don't have any particles. The

48 This would take a long time. Better buy two copies of this book to take with you.

concept of "space" would end when there are no more matter particles left to define it. What would be beyond that? Not empty space.

EXHIBIT B: SPACE

That is a pretty weird and counterintuitive way of thinking about space, especially given that we have never experienced the concept of non-space. But weird never stood in the way of physics, so keep an open mind.

Which Space Is the Place?

Which of these ideas about space is correct? Is space like an infinite void waiting to be filled? Or does it only exist in the context of matter?

It turns out that we are fairly certain that space is neither of these things. Space is definitely *not* an empty void and it is definitely *not* just a relationship between matter. We know this because we have seen space do things that fit neither of those ideas. We have observed space *bend and ripple and expand.*

This is the part where your brain goes, "Whaaaaat . . . ?"

If you are paying attention, you should be a little confused when you read the phrases "bending of space" and "expanding of space." What could that possibly mean? How does it make any sense? If space is an idea, then it can't be bent or expanded any more than it can be chopped into cubes and sautéed with cilantro.[49] If space is our ruler for measuring

49 Except in California. They can do anything with cilantro.

the location of stuff, how do you measure the bending or expanding of space?

Good questions! The reason this idea of space bending is so confusing is that most of us grow up with a mental picture of space as an invisible backdrop in which things happen. Maybe you imagine space to be like that theater stage we mentioned before, with hard wooden planks as a floor and rigid walls on all sides. And maybe you imagine that nothing in the universe could bend that stage because this abstract frame is not part of the universe but something that *contains* the universe.

Unfortunately, that is where your mental picture goes wrong. To make sense of general relativity and think about modern ideas of space, you have to give up the idea of space as an abstract stage and accept that it is a *physical thing*. You have to imagine that space has properties and behaviors, and that it reacts to the matter in the universe. You can pinch space, squeeze it, and, yes, even fill it with cilantro.[50]

At this point, your brain might be sounding "*what the #@#$?!?!*" nonsense alarms. Maybe you even threw this book against the wall and scoffed. That is totally understandable. Once you pick it back up, prepare to bear with us, because the real craziness is yet to come. Your nonsense alarms will be exhausted by the time we're done. But we need to unpack these concepts carefully to understand the ideas here and appreciate the truly strange and basic mysteries about space that remain unanswered.

Space Goo, You're Swimming in It

How can space be a physical thing that ripples and bends, and what does that mean?

50 Stay tuned for our follow-up book, *Cooking with Physicists*.

It means that instead of being like an empty room (a really big room) space is more like a huge blob of thick goo. Normally, things can move around in the goo without any problems, just like we can move around a room full of air without noticing all the air particles. But under certain circumstances, this goo can bend, changing the way that things move through it. It can also squish and make waves, changing the shape of the things inside it.

EXHIBIT C: SPACE

This goo (we'll call it "space goo") is not a perfect analogy for the nature of space, but it's an analogy that helps you imagine that the space you are sitting in right now at this moment is not necessarily fixed and abstract.[51] Instead, you are sitting in some concrete *thing*, and that thing can stretch or jiggle or distort in ways that you may not be perceiving.

Maybe a ripple of space just passed through you. Or maybe we are being stretched in an odd direction at this moment and don't even know it. In fact, we didn't even notice until recently that the goo did anything but sit there, goo-ing nowhere, which is why we confused it with nothingness.

So what can this space goo do? It turns out it can do a lot of weird things.

First, space can expand. Let's think carefully for a minute about what it means for space to expand. That means things get farther apart from each other *without actually moving through the goo*. In our analogy, imagine that you are sitting in the goo, and suddenly the goo started

51 Goo is not a perfect analogy because goo is a thing that exists inside space, whereas space has goo-like properties but we don't know if it exists inside anything else.

growing and expanding. If you were sitting across from another person, that person would now be farther away from you without either of you having moved relative to the goo.

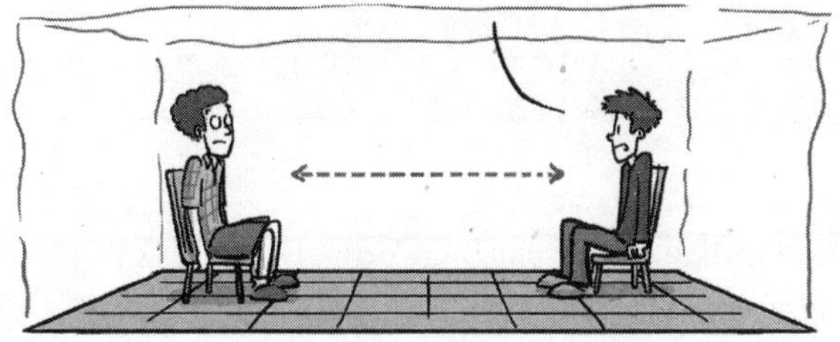
I FEEL LIKE WE'RE DRIFTING APART.

SPACE EXPANSION

How could we know that the goo expanded? Wouldn't a ruler we use to measure the goo *also* expand? It's true that the space between all the atoms in the ruler would expand, pulling them apart. And if our ruler was made out of extra-soft taffy, it would also expand. But if you use a rigid ruler, all of its atoms would hold on to one another tightly (with electromagnetic forces), and the ruler would stay the same length, allowing you to notice that more space was created.

And we know that space can expand because we have *seen* it expanding—this is how dark energy was discovered. We know that in the early universe space expanded and stretched at shocking rates, and that a similar expansion is still happening today. See chapter 14 for a discussion of the Big Bang (which blew up the early universe) and chapter 3 for a

MEASURING SPACE EXPANSION WITH:

discussion of dark energy, which is currently working to push us away from everything else in the universe.

We also know that space can *bend*. Our goo can be squished and deformed just like taffy can. We know this because in Einstein's theory of general relativity that's what gravity *is*: the bending of space.[52] When something has mass, it causes the space around it to distort and change shape.

When space changes shape, things no longer move through it the way you might first imagine. Rather than moving in a straight line, a baseball passing through a blob of bent goo will curve along with it. If the goo is severely distorted by something heavy, like a bowling ball, the baseball might even move in a loop around it—the same way the moon orbits the Earth, or the Earth orbits the Sun.

And this is something we can actually see with our naked eyes! Light, for example, bends its path when it passes near massive objects like our

52 Einstein famously didn't say "Goo does not play dice."

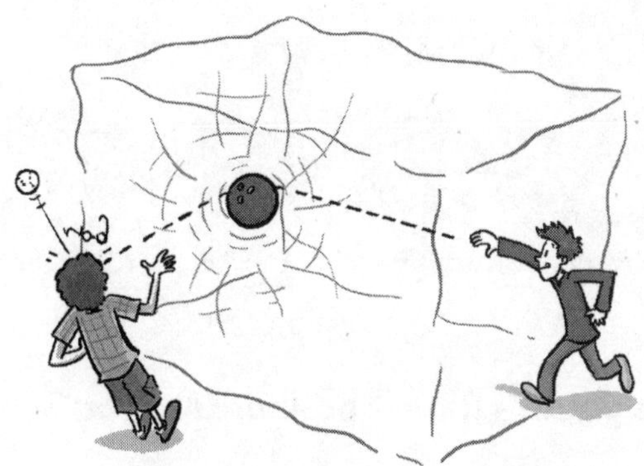

THE EINSTEIN TRICK SHOT.

Sun or giant blobs of dark matter. If gravity was just a force between objects with mass—rather than the bending of space—then it shouldn't be able to pull on photons, which have no mass. The only way to explain how light's path can be bent is if it's the space itself that is bending.

Finally, we know that space can *ripple*. This is not too far-fetched given that we know that space can stretch and bend. But what is interesting is that the stretching and bending can *propagate* across our space goo; this is called a gravitational wave. If you cause a sudden distortion of space, that distortion will radiate outward like a sound wave or a ripple inside of a liquid. This kind of behavior could only happen if space has a certain physical nature to it and is not just an abstract concept or pure emptiness.

We know this rippling behavior is real because (a) general relativity predicts these ripples, and (b) we have actually sensed these ripples. Somewhere in the universe, two massive black holes were locked in a frenzied spin around each other, and as they spun, they caused huge distortions in space that radiated outward into space. Using very sensitive equipment, we detected those space ripples here on Earth.

You can think of these ripples as waves of space stretching and compressing. Actually, when a space ripple passes through, space shrinks in one direction and expands in another direction.

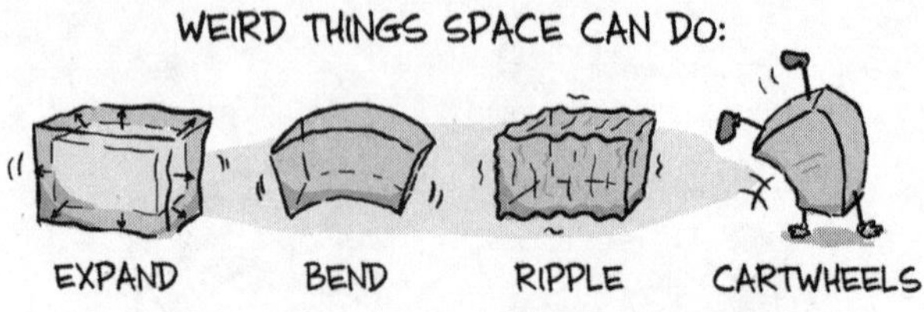

This Sounds Ridic-goo-lous. Are You Sure?

As crazy as it may sound that space is a thing and not just pure emptiness, this is what our experience of the universe tells us. Our experimental observations make it pretty clear that the distance between objects in space is not measured on an invisible abstract backdrop but depends on the properties of the space goo in which we all live, eat cookies, and chop cilantro.

But while thinking of space as a dynamic thing with physical properties and behaviors might explain weird phenomena like space bending and stretching, it only leads to more questions.

For example, you might be tempted to say that what we used to call space should now be called physics goo ("phgoo") but that this goo has to be *in* something, which we could now call space again. That would be clever, but as far as we know (which to date is not very far), the goo does not need to be in anything else. When it bends and curves, this is *intrinsic bending* that changes the relationships between parts of space, not the bending of the goo relative to some larger room that it fills.

But just because our space goo doesn't *need* to sit inside of something else doesn't mean that it is *not* sitting inside something else. Perhaps what we call space is actually sitting inside some larger "superspace."[53] And perhaps that superspace *is* like an infinite emptiness, but we have no idea.

Is it possible to have parts of the universe without space? In other

53 Suspiciously, superspace has never been seen in the same room as regular mild-mannered-reporter space.

words, if space is a goo, is it possible for there to be not-goo, or the absence of goo? The meaning of those concepts is not very clear because all of our physical laws assume the existence of space, so what laws could operate outside of space? We have no idea.

The fact is that this new understanding of space as a thing has come recently, and we are at the very beginning of understanding what space is. In many ways, we are still hobbled by our intuitive notions. These notions served us well when early men and women were hunting for game and foraging for prehistoric cilantro, but we need to break the shackles of these concepts and realize that space is very different from what we imagined.

Straight Thinking about Bent Space

If your brain is not yet hurting from all these gooey space-bending concepts, here is another mystery about space: Is space flat or curved (and if it's curved, which way does it curve)?

These are crazy questions, but they are not that hard to ask once you accept the notion that space is malleable. If space can bend around objects with mass, could it have an overall curvature to it? It's like asking if our goo is flat: You know that it can jiggle and deform if you push any point on it, but does it sag overall? Or does it sit perfectly straight? You can ask these questions about space, too.

Answering these questions about space would have an enormous impact on our notion of the universe. For example, if space is flat, it means that if you travel in one direction forever you could just keep going, possibly to infinity.

But if space is curved, then other interesting things might happen. If space has an overall positive curvature, then going off in one direction might actually make you loop around and come back to the same spot from the opposite direction! This is useful information if, for example, you don't like the idea of people sneaking up behind you.

THE LONGEST PRACTICAL JOKE IN THE UNIVERSE

Explaining the idea of curved space is very difficult because our brains are simply not well equipped to visualize concepts like these. Why would they be? Most of our everyday experience (like evading predators or finding our keys) deals with a three-dimensional world that seems pretty fixed (although if we are ever attacked by advanced aliens that can manipulate the curvature of space, we hope we, too, can figure it out quickly).

What would it mean for space to have a curvature? One way to visualize it is to pretend for a second that we live in a two-dimensional world, like being trapped in a sheet of paper. That means we can only move in two directions. Now, if that sheet we live in lies perfectly straight, we say that our space is flat.

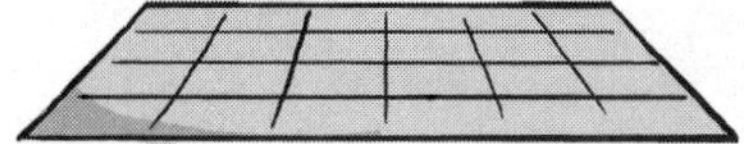

But if for some reason that sheet of paper is bent, then we say that the space is curved.

And there are two ways that the paper can be bent. It can all be curved in one direction (called "positive curvature") or it can be bent in different directions like a horse saddle or a Pringles potato chip (this is called "negative curvature" or "breaking your diet").

Here is the cool part: if we find out that space is flat everywhere, it means that the sheet of paper (space) could potentially go on forever. But if we find out that space has a positive curvature everywhere, then there's only one shape that has positive curvature everywhere: a sphere. Or to be more technical, a spheroid (i.e., a potato). This is one way in which our universe could loop around itself. We could all be living in the three-dimensional equivalent of a potato, which means that no matter which direction you go you end up coming back around to the same spot.

So which is it? Is our space flat or does it have an overall curvature? And if you live in an apartment, does it mean your flat is flat-out not flat, flatly speaking?

Well, in this case, it turns out that we do have an answer, which is that

space does appear to be "pretty flat," as in space is within 0.4 percent of being flat. Scientists, through two very different methods, have calculated that the curvature of space (at least the space we can see) is very nearly zero.

What are these two ways? One of the ways is by measuring triangles. An interesting thing about curvature is that triangles in a curved space don't follow the same rules as triangles in flat space. Think back to our sheet-of-paper analogy. A triangle drawn on a flat sheet of paper is going to look different than a triangle drawn on a curved surface.

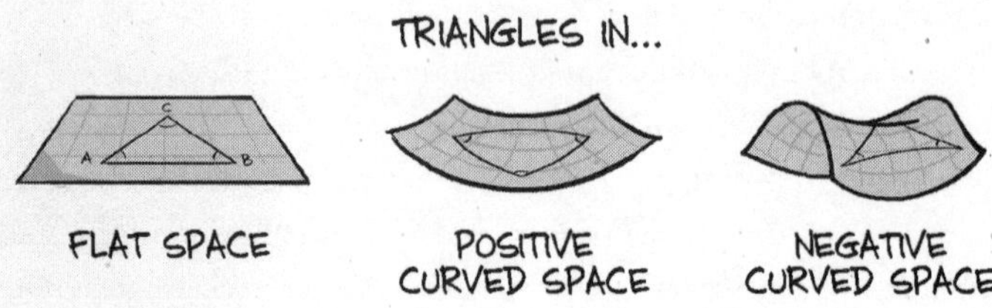

Scientists have done the equivalent of measuring triangles drawn in our three-dimensional universe by looking at a picture of the early universe (remember the cosmic microwave background from chapter 3?) and studying the spatial relationship between different points on that picture. And what they found was that the triangles they measured correspond to those of flat space.

The other way in which we know that space is basically flat is by looking at the thing that causes space to curve in the first place: the energy in the universe. According to general relativity, there is a specific amount of energy in the universe (energy density, actually) that will cause space to bend in one direction or the other. It turns out that the amount of energy density that we can measure in our universe is exactly the right amount needed to cause the space that we can see to not bend at all (within a margin of error of 0.4 percent).

Some of you might be disappointed to learn we don't live in a cool three-dimensional cosmic potato that loops around if you go in one direction forever. Sure, who hasn't dreamed of doing Evel Knievel–style spins around the entire universe on a rocket motorcycle? But instead of feeling disappointed by the fact that we live in a boring flat universe, you might want to be a little intrigued. Why? Because as far as we know, the fact that we live in a flat universe is a gigantic cosmic-level coincidence.

Think about it. All the mass and energy in the universe is what gives space its curvature (remember that mass and energy distort space), and if we had just a little bit more mass and energy than we have right now, space would have curved one way. And if we had just a little bit less than we have right now, space would have curved the other way. But we seem to have *just* the right amount to make space perfectly flat as far as we can tell. In fact, the exact amount is about five hydrogen atoms per cubic meter of space. If we had had *six* hydrogen atoms per cubic meter of space, or *four*, the entire universe would have been a lot different (curvier and sexier, but different).

And it gets stranger. Since the curvature of space affects the motion of matter, and matter affects the curvature of space, there are feedback effects. This means that if there had been just a little too much matter or not quite enough matter in the early days of the universe, so that we weren't right at this critical density to make space flat, then it would have pushed things even farther from flat. For space to be pretty flat now means that it had to be *extremely* flat in the early universe, *or* there has to be something else keeping it flat.

This is one of the biggest mysteries about space. Not only do we not know what exactly space is, we also don't know why it happens to be the way it is. Our knowledge in this matter appears to fall . . . flat.

The Shape of Space

The curvature of space is not the only thing we have deep questions about when it comes to the nature of space. Once you accept that space is not an infinite void but rather a maybe-infinite physical thing with properties, you can ask all kinds of strange questions about it. For example, what is the size and shape of space?

The size and shape of space tell us how much space there is and how it is connected to itself. You might think that since space is flat, and not shaped like a potato or a horse saddle (or a potato on a horse saddle), the idea of the size and shape of space makes no sense. After all, if space is flat, it means that it must go on forever, right? Not necessarily!

DEFINITELY NOT THE SHAPE OF SPACE.

Space can be flat and infinite. Or it could be flat and have an edge to it. Or, even stranger, it could be flat and *still* loop around itself.

How can space have an edge? Actually, there's no reason why space can't have a boundary even if it is flat. For example, a disc is a flat two-dimensional surface with a smooth continuous edge. Perhaps three-dimensional space also has a boundary at some point thanks to some strange geometric property at its edges.

Even more intriguing is the possibility that space can be flat and still loop around itself. It would be like playing one of those video games (like *Asteroids* or *Pac-Man*) where if you move beyond the edge of the screen you simply appear on the other side. Space might be able to connect with itself in some way that we are not completely aware of yet. For example, wormholes are theoretical predictions of general relativity. In a wormhole, two different points in space that are far apart can be connected to each other. What if the edges of space are all connected together in a similar way? We have no idea.

Quantum Space

Finally, you can ask whether space is actually made up of tiny discrete bits of space, like the pixels on a TV screen, or infinitely smooth, such that there are an infinite number of places you can be between two points in space?

Scientists in ancient times might not have imagined that air is made up of tiny discrete molecules. After all, air appears to be continuous. It acts to fill any volume and it has interesting dynamical properties (like wind and weather). Yet we know that all these things we love about air (how it brushes gently against your cheek in a cool summer breeze or how it keeps us from asphyxiating) are actually the combined behavior of billions of individual air molecules, not the fundamental properties of the individual molecules themselves.

The smooth space scenario would appear to make more sense to us. After all, our experience of moving through space is that we glide through it in an easy, continuous way. We don't jump from pixel to pixel in a jerky fashion the way a video-game character does when it moves across the screen.

Or do we?

Given our current understanding of the universe, it would actually be *more* surprising if space turned out to be infinitely smooth. That's because we know that everything else is quantized. Matter is quantized, energy is quantized, forces are quantized, Girl Scout cookies are quantized. Moreover, quantum physics suggests that there might be a smallest distance that even makes sense, which is about 10^{-35} meters.[54] So from a quantum mechanical perspective, it would make sense if space was quantized. But, again, we really have no idea.

But having no idea hasn't stopped physicists from imagining crazy possibilities! If space *is* quantized, that means that when we move across space we are actually jumping from small little locations to other small little locations. In this view, space is a network of connected nodes, like the stations in a subway system. Each node represents a location, and the connections between nodes represent the relationships between these locations (i.e., which one is next to which other one). This is different from the idea that space is just the relationship between matter, because these nodes of space can be empty and still exist.

54 This length is not a made-up number even though it's hard to think about. It's the Planck length, our current best estimate for the smallest meaningful unit of distance. See chapter 16 for a detailed discussion.

Interestingly enough, these nodes *would not need to sit inside a larger space* or framework. They could just . . . be. In this scenario, what we call space would just be the relationships between the nodes, and all the particles in the universe would just be properties of this space rather than elements in it. For example, they might be vibrational modes of these nodes.

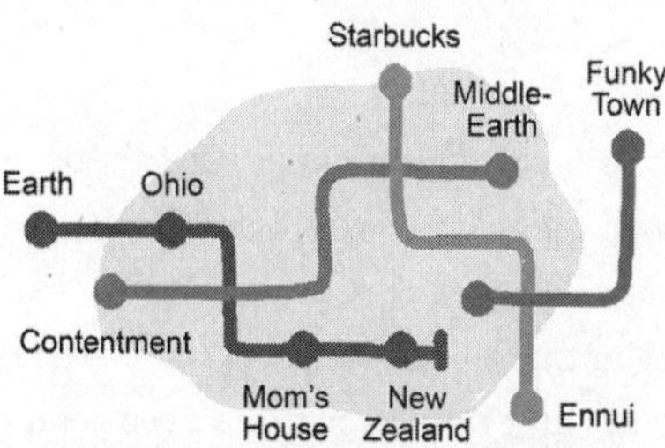

A MAP OF NODAL SPACE

This is not as far-fetched as it sounds. The current theory of particles is based on quantum fields that fill all of space. A field just means there is a number, or a value, associated with every point in that space. In this view, particles are just excited states of these fields. So we are not *too* far from this kind of theory already.

By the way, physicists love this type of idea, where something that seems fundamental to us (like space) comes out accidentally from something deeper. It gives them the sense that we have peeked behind the curtain to discover a deeper layer of reality. Some even suspect that the relationships between nodes of space are formed by the quantum entanglement of particles, but this is mathematical speculation by a bunch of overcaffeinated theorists.

The Mysteries of Space

To summarize, here are the major unresolved mysteries about space thus far:

- Space is a thing, but what is that thing?
- Is the space that we know all there is, or does it sit inside some larger metaspace?
- Are there parts of the universe that don't have space?
- Why is space flat?
- Is space quantized?
- Why does Anna from accounting not respect other people's personal space?

If you have read this far and either understood it deeply or just turned your nonsense alarm to mute, then we should not hesitate to explore the craziest concept about space (yes, it gets crazier).

If space is a physical thing—not a backdrop or frame—with dynamic properties such as twists and ripples, perhaps even built out of quantized bits of space, then we have to wonder: What *else* can space do?

Like air, perhaps it has different states and phases. Under extreme conditions, maybe it can arrange itself in very unexpected ways or have weird unexpected properties in the same way that air behaves differently whether it's in liquid, gas, or solid form. Perhaps the space we know and love and occupy (sometimes more than we'd like) is only one rare type of space and there are other types of space out there in the universe just waiting for us to figure out how to create and manipulate them.

The most intriguing tool we have to answer this question is the fact that space is distorted by mass and energy. In order to understand what space is and what it can do, our best bet is to push it to extremes by looking carefully at places where cosmically huge masses are squeezing and straining it: black holes. If we could explore near black holes, we might see space shredded and chopped in ways that cause our nonsense alarms to explode.

And the exciting thing is that we are closer than ever to being able to probe these extreme deformations of space. Whereas before we were deaf to the ripples of gravitational waves moving through the universe, we now have the ability to listen in to the cosmic events that are shaking and disturbing the goo of space. Perhaps in the near future we will understand more about the exact nature of space and get at these deep questions that are literally all around us.

So don't space out. And save some space in your brain for the answers.

8.

What Is Time?

In Which We Learn That Time Is of the (Unknown) Essence

We have seen that basic concepts like space and mass and matter turn out to be much more mysterious than you probably thought. What other basic elements of our world might be hiding their strangeness in plain sight? It's time to ask that timely question:

What is time?

FINALLY! IT'S ABOUT TIME!
THIS BOOK

If you were an alien visitor to Earth trying to learn our language by eavesdropping on conversations in cafés and grocery stores, you might have a hard time answering this question. Humans spend a lot of time talking about time but almost no time talking about what time actually is!

We check the time all the time. We talk about having bad times, good times, old times, crazy times. We save time, keep time, make time, spend time, cut time, pass time. Time can be up, out, over, and even down. It waits for no man or woman! Sometimes it flies and sometimes it creeps up on you and sometimes it ticks away. Most of the time, we simply run out of time.

But what is it? Is it a physical thing (like matter or space) or an abstract concept we layer on top of our experience of the universe?

If you were hoping physicists had an answer to the deep and somewhat confusing question of time, this is not the right time for that. Time is still one of the great mysteries in physics, one that calls into question the very definition of what physics is. So let's take our time and carefully explore this timeless topic.

TOO MANY TIME PUNS?
GIVE IT TIME.

A Definition of Time

Of all the questions you can ask about the universe, the ones that are the *most* fun are the ones that sound simple but are actually very hard to answer. They're the kind of questions that make you scratch your head and realize that there are basic things staring us right in the face that we don't have clear explanations for.

This kind of question raises the possibility that we could be looking at things all wrong, as we have done in the past (e.g., "The Earth is flat" or "Hey, let me put some leeches on you to cure your disease!"), and that getting a firm, concrete answer could change the way we think about the universe and our place in it. The stakes are very high!

The first thing we should do is try to define what time is. After all, this is how physics approaches difficult questions. First, we come up with a careful definition of the thing you are trying to understand, and then we follow it with a mathematical description that lets you apply the power of logic and experiment to guide you the rest of the way.

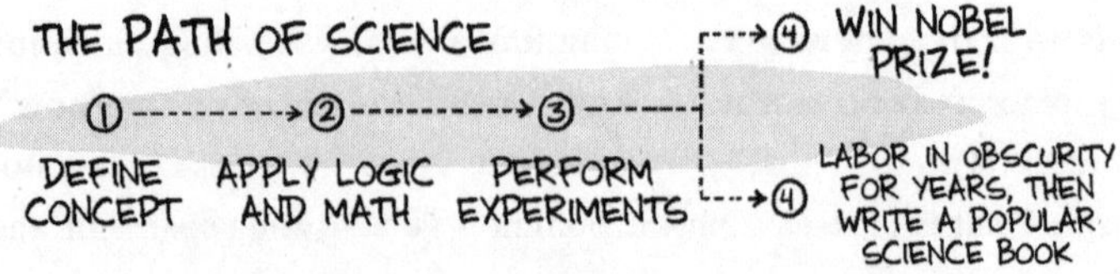
THE PATH OF SCIENCE
① DEFINE CONCEPT
② APPLY LOGIC AND MATH
③ PERFORM EXPERIMENTS
④ WIN NOBEL PRIZE!
④ LABOR IN OBSCURITY FOR YEARS, THEN WRITE A POPULAR SCIENCE BOOK

So what *is* time? If you stopped random strangers on the street today and asked them to define time, you might get answers like:

"Time is the difference between *then* and *now.*"
"Time is what tells us when things happen."
"Time is what clocks measure."
"Time is money so leave me alone!"

These are plausible definitions of time, but they all raise even more questions. For example, you can ask, "Why is there a 'then' and a 'now' in the first place?" or "What does 'when' even mean?" or "Aren't clocks subject to time also?" or "Who has time for all this?"

LITERARY DEFINITIONS OF TIME

THAT WHICH CAN SIMULTANEOUSLY BE THE BEST OF AND THE WORST OF

A BIG BALL OF WIBBLY WOBBLY TIMEY WIMEY STUFF

HAMMER TIME

It seems difficult to make progress if we can't even describe time, but there's no reason for alarm. Though the question "What is time?" sounds like something a five-year-old would ask, it would not be the first time that we have trouble defining or precisely describing something we are very familiar with.[55] It happens in other fields, too; biologists have been arguing for decades about the definition of "life" (the zombie-rights activists are a powerful lobby group), neuroscientists bicker over the definition of "consciousness," and Godzillalogists[56] can't agree on the definition of "monster."

Part of the difficulty in defining time is that it is so ingrained in our experience and our way of thinking. Time is how we relate the "now" we have now with the "now" we had before. Whatever we are feeling now is what we call the present, but the present is fleeting and ephemeral: there's

55 Physicists: five-year-olds who never grew up.

56 Sorry, kids, not a real job.

no way to savor it or stretch it as you might a tasty bite of chocolate cake. Every moment slips immediately from the intense experience of the present to a fading memory of the past.

THE UNBEARABLE LIGHTNESS OF NOWNESS

But time is also about the future. Being able to connect the future to the past and the present is very important. If you are a caveperson hoping to survive the next winter or a modern person who needs a place to charge her smartphone, thinking about the future and extrapolating from the past is absolutely critical to your survival. So it's hard to imagine the human experience without the concept of time.

The same is true for the way that physics thinks about time. In fact, time is embedded in the very definition of physics! Physics, on good authority (Wikipedia), is nothing more than "the study of matter and its motion through space and time." Even the word "motion" assumes the concept of time. The basic job of physics is to use the past to understand what futures are possible and how we could affect them. Physics makes no sense without time.

WHAT DO YOU CALL A PHYSICIST WITH TOO MUCH TIME?
A PROFESSOR.

The truth is that any definition of time by humans is likely to be distorted by the nature of our experience. Think about it: just thinking about time *requires* time! It might be that alien physicists don't have the same concept of time that we do because their experience and patterns of thought are different in some deeply alien way that our current subjective experience prevents us from truly grasping.

So Tell Us Already: What Is Time?

Let's talk about ferrets.

To get a better grip on how physicists think about time, let's consider a common scenario. For example, suppose your pet ferrets are planning to drop a water balloon on your head when you get home from work. Happens all the time, right?

Now, instead of thinking about time as a smooth stream of experience, chop it into slices and imagine that it works the way a movie does: by stringing together many static snapshots.

For physicists, each of these snapshots describes the state of affairs at some moment. So you have a series of snapshots:

1. You walk innocently to your front door, whistling and carefree.
2. The ferrets nudge the water balloon into position.
3. You put your key in the front door.
4. The ferrets launch the payload.
5. You are drenched.
6. The ferrets laugh and laugh.

Each snapshot is a description of the local situation: where everything is and what it is doing at one moment. Each is frozen, static, without change. If we didn't have the concept of time, the universe would be one of these frozen snapshots incapable of change or motion.

Fortunately, our universe is more interesting than that. These snapshots don't exist independently of one another. Time relates them to each other in two important ways.

First, it links the snapshots together in a chain, putting them in a particular order. For example, this sequence wouldn't feel right to us if it was ordered differently.

Second, it requires that the snapshots be causally connected to one another. That means that each moment in the universe depends on what happened just before that. This is nothing more than cause and effect. For example, you can't be on your couch eating ice cream one moment and then be halfway through a marathon in the next.

This is precisely the job of the laws of physics: to tell us how the universe can change and how it cannot. Given a specific snapshot, physics tells us what future snapshots are possible, which are likely, and which cannot happen. And time is the basis of these requirements. Without time, we have to imagine a static universe because any sort of change or motion requires time.

IN A TIMELESS UNIVERSE, YOU NEVER FIND OUT WHAT HAPPENS NEXT.

So how do we relate this back to our smooth experience of time? Well, we can stitch these snapshots together to make a movie that is as smooth and continuous as we like by making the time separation between slices as small as we like.[57]

This is exactly what our mathematical language for physics—calculus—was invented to do: convert many tiny little slices into a smooth variation. When you are watching a movie, you don't notice that it is actually a sequence of frozen images because the time separation is very small. In the same way, our description of a universe full of change and motion is a set of ordered, static snapshots related to one another by the laws of physics. Time is the ordering and spacing of those snapshots.

TIME = SCRAPBOOKING

I Am Still a Little Confused!

If the preceding definition of time seemed a little fuzzy and unsatisfying, then take a number. Physicists, philosophers, and five-year-olds have been arguing for centuries over what exactly time is. To date, there is no universally agreed-upon set of words to describe time.[58] If you open any textbook on physics, few will even try to tackle the subject. This is one of the central mysteries of time: that it defies an exact definition. It is so ingrained in how we view the world and in our tools for understanding that world that the best we can do is to talk in general terms and try to distract you with fancy words like "calculus" and "ferrets."

Our entire apparatus for understanding our place in the universe assumes this continuous experience of time, and for the most part, it actually works.[59] But even so, there are still many questions we can ask about this fuzzy concept of time. For example, why do we have it at all? Why

57 Almost. The uncertainty principle also applies to time, so there is some basic fuzziness.

58 To be fair, there is probably also no universally agreed-upon set of words to describe *anything*.

59 At least for the 5 percent of the universe we are familiar with.

does it seem to move only forward? Does it, in fact, move only forward? Some say it is part of space-time, but why is it so different from space? Can we go back in time and buy Google stock in 2001?

It is time to go deeper into time.

Time Is the Fourth Dimension (or Is It?)

You might have noticed that the idea of time as a long continuum that we can travel along bears a striking similarity to another fundamental piece of the universe: space.

The same logic of slicing our journey through time into static snapshots can also be applied to our motion through space. This leads us to consider the possibility that time and space are closely related.

Indeed, modern physics tells us that time and space *are* very similar, and in many ways, it is totally correct to think of time as another direction in which we can move. Let that idea settle in for a minute. As is often the case, it is easier to think about if you simplify our universe. Imagine that there was only one direction you could move in space rather than the three we are familiar with.

Now imagine a day in the life of your one-dimensional pet ferret. He wakes up in the morning, and he has a lot of work to do (those water-balloon practical jokes don't plan and organize themselves!). Let's imagine that he makes several trips back and forth to the balloon store before your return.

The plot shows the ferret moving along that one dimension through-

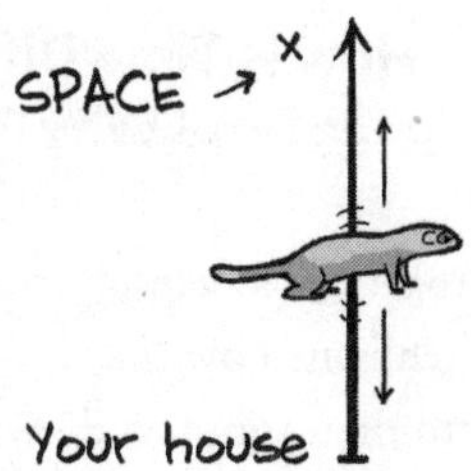

out the day. But you can also think of your ferret's path through a two-dimensional plane called space-time. In fact, in physics, the mathematics of motion is simpler and cleaner if you treat time as the fourth dimension (assuming we have only three spatial dimensions. See chapter 9 for the other possibilities).

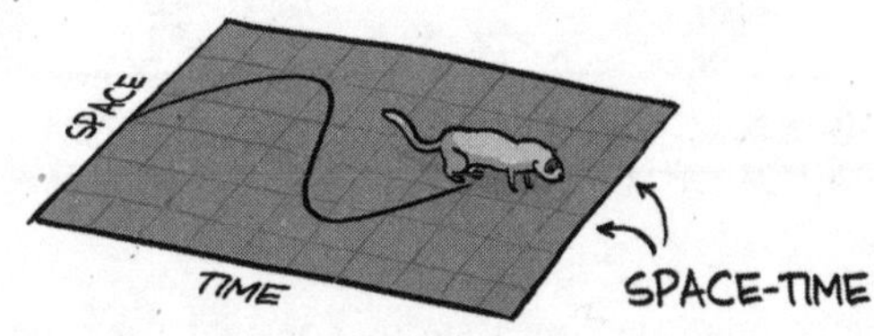

It is always very satisfying to connect two different concepts together and realize that they're part of some larger framework. This is often the first step in gaining some deep understanding. Like when you realize that chocolate and peanut butter taste so good together, they must be part of some deep universal chocolate-peanut continuum.

But don't get too excited. This connection between space and time doesn't mean that you can think of time as a dimension of space with all of the implications that come along with it. There are several ways that time is very different from space. These are some of the remaining basic mysteries of time, and we hope that they give us the clues to understanding the bigger picture of space-time. So far, we barely know how to ask the questions.

Question #1: How Is Time Different from Space (and Why)?

Connecting time and space together is helpful because it shows us how they are similar, but it also highlights how they are different. You have a very different relationship with time than you do with space.

For starters, you are free to move around in space however you like. You can walk in circles or go backward to places you have already been to. You can also move around space at whatever speed you like, either fast or slow. Or you can sit in one place and not move at all for a while. But time is different. You have no such freedom with time.

You move through time at a steady constant pace (one second per second to be precise).[60] You can't backtrack or do loops in time. You can't suddenly decide to go backward in time and be in a different spot in space than you were at that previous time. Even though you can be in the same position in space at different times, you can't be in different positions in space at the same time.

Just as weird: it's normal to think of something having a fixed location (one position in space), but it would be really bizarre to have a fixed time. This is because time marches on like a wavefront. Once a moment is gone, it is gone forever (like those Girl Scout cookies that were sitting on the counter). In contrast, your location in space is variable and unconstrained. There are plenty of places in space that you will never visit

60 If you are near black holes or moving at high speeds, your time can go slower or faster for other people, but you still experience one second per second.

during your life and plenty that you will visit multiple times. But between the moments of your birth and your death, you move in only one direction through time. Unless your life story is very peculiar (such as living on a colonization ship taking a generations-long journey between galaxies) your trip through time will be much different than your trip through space.

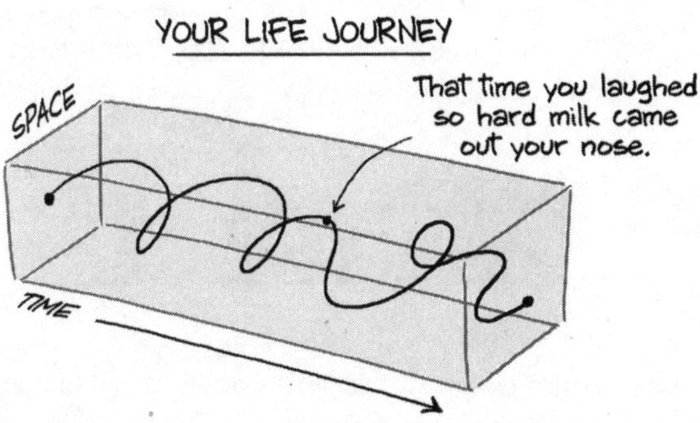

While thinking about time as another dimension is mathematically convenient in our theories, it's important to keep in mind that there are significant differences that make time unique. Time works differently than space does because time is not a set of interconnected locations. Rather, we think of time as a linking together of causally connected static snapshots of the universe, and this has enormous consequences on what we can (and cannot) do with time.

Question #2: Can We Travel Back in Time?

The lessons from this book should make you very skeptical of thinking that anything is impossible. After all, maybe what we say is impossible now will change once we gain a better understanding of the universe. Many things that seemed impossible are now commonplace, like having access to most of human knowledge and inane trivia using a pocket-size phone device.[61]

But in the case of time travel, modern physics is as certain as it can be that this is not possible. Any scenario in which you can travel backward in time quickly leads to paradoxes that violate deep and basic assumptions about the workings of the universe.

WHAT KIND OF DOCTOR ARE YOU, "DOC" BROWN?

In some science fiction stories, aliens or advanced humans are able to view time as a spatial dimension and move back and forth over it; this allows them to move through time the way you and I walk up and down the hallway. And while these are very fun stories to read and enjoy, they have serious problems from a physics perspective.[62]

First, moving backward in time can break causality. If you want the universe to make sense, then that is a *big deal*. If you don't mind if effects happen before causes (your credit card is billed before you buy this book, or your ferrets eat your breakfast before you prepare it), then you are more open-minded than we are.

Without causality, nothing really makes sense. For example, if your ferrets get tired of dropping water balloons on your head when you come home because you've grown wary and anticipate it, they might build a time machine to travel back to an earlier date in 2005 *before* you had ferrets, when you were still naïve and easy to startle. If they succeed in their plan to give you a dousing, it could have unintended consequences. What if you were on the fence about getting ferrets in the first place and

61 Still impossible: getting good reception when you really need it.

62 Physics: ruining fun since ancient times.

this helped make up your mind? If you decide not to get ferrets, then there are no ferrets to later douse you until they become bored and build a time machine! This, in turn, means no 2005 dousing, which leads you to get ferrets, etc. You are trapped in an eternal cycle of ferret inconsistency. The moral of the story is that time travel is not possible because it violates causality and that you should think twice before you decide to get ferrets. This is the famous ferret paradox.[63]

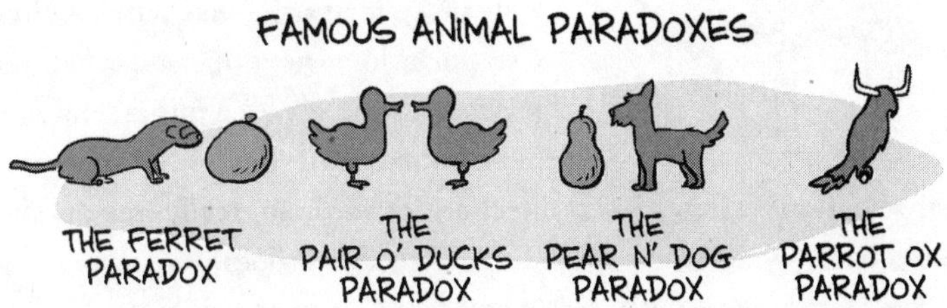

More important, think carefully about what's happening in these fun science fiction stories. The aliens are *moving* through this fictional space-time; but, remember, *motion* implies time. These aliens have some location in space-time and then later they have another location. What does "later" mean? These well-meaning authors have reinserted the notion of linear time on top of their space-time universe. The lesson is that it's hard to come up with a consistent universe (even a fictional one) in which time is more like space.

Question #3: Why Does Time Move Forward?

Since we can't move backward in time, you might reasonably ask, "Why does time move forward?"

The concept of time *not* moving forward is bizarre to us. You wouldn't expect that an oven would turn cooked food raw or that ice cubes might form in your drink on a hot day or that those Girl Scout cookies would

63 Famous according to us.

uneat themselves. All of these things are very familiar in the forward direction but would make you want to dial down the dose on your medication if you saw them happen in reverse.

In the same way, you can remember things that happen in the past, but you can't remember things that happen in the future.[64] Time appears to have a preferred direction, and we have *no idea why*.

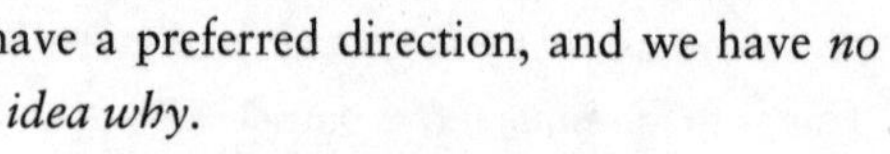

This basic question—why does time only move forward?—has puzzled physicists for a long time. In fact, what does "forward in time" even mean? In some universe where time flows the other way, their scientists might call *that* direction forward. So, really, the question should be: Why does time move in the direction it does?

The first thing to consider is whether the universe would even work if time went the other way. Do the laws of physics require that time flow in one direction? Imagine that you are watching a video recording of some universe. Could you tell by careful examination whether the video was being played forward or backward? For example, let's say you're watching a video of a ball bouncing up and down. As long as the ball bounces perfectly (and doesn't lose any energy to friction or air resistance) then the forward and backward versions of this video would look *exactly the same*! The same is true for particles of gas bouncing inside a canister or molecules of water flowing in a river. Even quantum mechanics works just

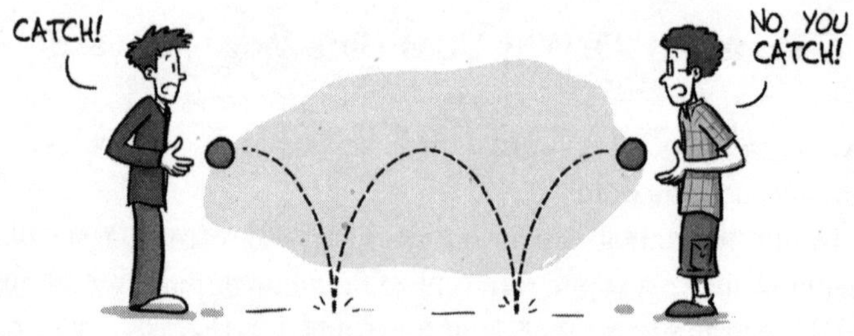

64 If you remember the future, give us a call. We have questions for you.

fine backward.[65] In fact, almost every law of physics would work just as well going forward as in reverse.

But not all of them.

The example of a perfectly bouncing ball is not realistic because it ignores the friction of the ball on the ground and air resistance and lots of other ways in which the energy of the ball is dissipated into heat. After a few bounces, even your pet ferret's favorite Super Ball will stop bouncing quite as high and eventually settle on the ground. All of its energy will be converted into heat of the air molecules or the ball molecules or the ground molecules.

Now imagine how bizarre a video of a real bouncing ball would look in reverse: a ball sitting on the ground would suddenly start bouncing higher and higher. The energy flow would look even stranger: the air and ball and ground would cool down a bit, and the lost heat would be converted into the motion of the ball.

You could definitely tell the difference between forward and backward in this example. The same is true for the other examples above: cooking food, melting ice cubes, and eating cookies. But if most of the laws of physics work just fine in reverse—especially the microphysics of heat and diffusion—why do these macroscopic processes seem to happen in only one direction? The reason is the amount of disorder in the system, known as entropy, which has a very strong preference for one direction in time.

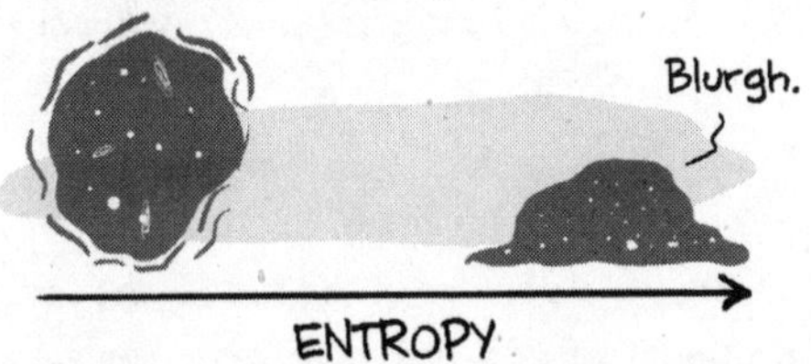

Entropy always increases with time. This is known as the second law of thermodynamics. Think of entropy as the amount of disorder in something. When you forget to feed your ferret, and it trashes your living room

65 Except for the collapsing of the wave function, which some argue is irreversible and others argue is a loss of coherence. Others just argue for argument's sake.

and knocks over your perfect stack of signed copies of this book, it has increased the entropy of your living room by increasing the disorder.

If you come home and reorganize it, then you decrease the entropy of the living room, but doing so takes quite a bit of energy, which you release as heat and frustration and muttering under your breath about how you told your roommate that a ferret was a bad idea. That energy you release while organizing your living room keeps the total entropy increasing. Whenever you create any sort of localized order—stacking books, making marks on a sheet of paper, or running your air-conditioning, you are simultaneously creating disorder as a by-product, usually as heat. According to the second law, it is impossible, *on average*, to decrease the total entropy in the forward direction of time.

(Note: this is a probabilistic statement. Technically, it is possible for a mob of angry ferrets to accidentally organize themselves into a perfectly ordered posse, thereby decreasing their entropy, but this is a tiny probability. Isolated accidents are allowed, but on average entropy always increases.)

This has some chilling consequences: because entropy only increases, eventually, very, very, very, very far in the future, the universe will reach some maximum amount of disorder, which goes by the cozy sounding name of "the heat death of the universe." In this state, the whole universe will be at the same temperature, which means everything will be completely disordered, with no little useful pockets of ordered structure (like humans). Until then, creating local pockets of order by making compensating pockets of disorder is only possible because the universe has not yet reached maximum disorder, so there is still wiggle room.

Now think backward in time. At every moment in the past, the universe had *less* entropy (more order) than it does now, all the way back to the moment of the Big Bang. Think of the Big Bang as the moment before your moving trucks and small children arrive at your pristine new house. This initial condition of the universe, when entropy was lowest, determines how much time there is between the birth and heat death of the universe. If the universe had begun with a huge amount of disorder, there would not be much time left before the heat death. In our case, it appears that the universe started as very highly ordered, giving us a lot of time before we get to maximum entropy.

Why did the universe start in such a highly organized low-entropy configuration in the first place? We have no idea. But we sure are lucky that it did, since it left plenty of time between the start and the finish to do interesting things, such as make planets and people and popsicles.

Does Entropy Help Us Understand Time?

Entropy is one of the few physical laws that cares one way or the other about how time flows.

Most of the processes that affect entropy, such as the laws of kinematics that affect how gas molecules bounce off one another, could work perfectly well going backward. But in aggregate, they follow a law that requires the amount of order to decrease with time. So time and entropy are connected in some way. But so far, we only have a correlation: entropy increases with time.

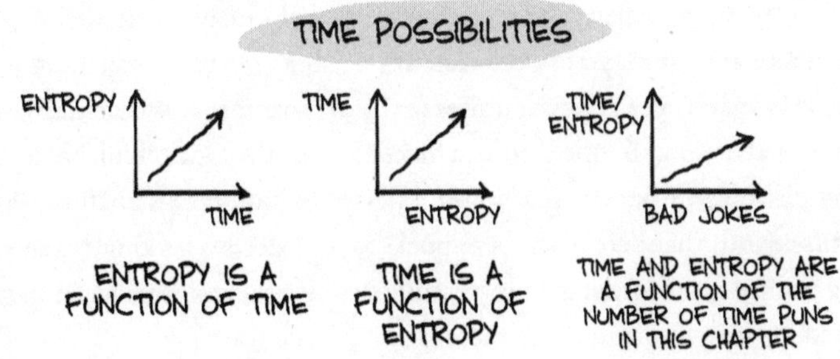

Does that mean that entropy *causes* time to go only forward, the way a hill causes water to flow only down? Or does entropy *follow* the arrow of time like debris caught in a tornado?

Even if you accept that entropy increases with time, it still doesn't explain why time goes only forward. For example, you can imagine a universe where time goes backward and entropy *decreases* with *negative* time, which would maintain the relationship *and* not violate the second law of thermodynamics!

So entropy is not so much an insight as much as it is a clue. It is one of the few clues we have about how time works, so it deserves our careful attention. Is entropy the key to understanding the direction of time? Though many may speculate, we still don't know. More than that, we have very few ways of figuring this out.

Time and Particles

When it comes to little particles, they seem to be generally ambivalent about the direction of time. For example, an electron is happy to radiate a photon or absorb one. Two quarks can fuse together to make a Z boson, or a Z boson can decay into two quarks. For the most part, you could not tell which direction time flows in our universe by watching individual particles interact. But not in every case. There is one kind of particle interaction that works differently if you run time forward or backward.

The weak force, the one responsible for nuclear decay and mediated by the W and Z bosons, has a part that prefers one direction. The details are not terribly important to understand and the effect is small, but it is real. For example, when a pair of quarks are held together by the strong force, there are sometimes two different possible arrangements. They can switch back and forth between these two arrangements using the weak force, but switching in one direction takes longer than switching back. So playing a video of this process in reverse would look different than playing it forward.

What does this have to do with time? We don't know precisely, but it smells like a useful clue.

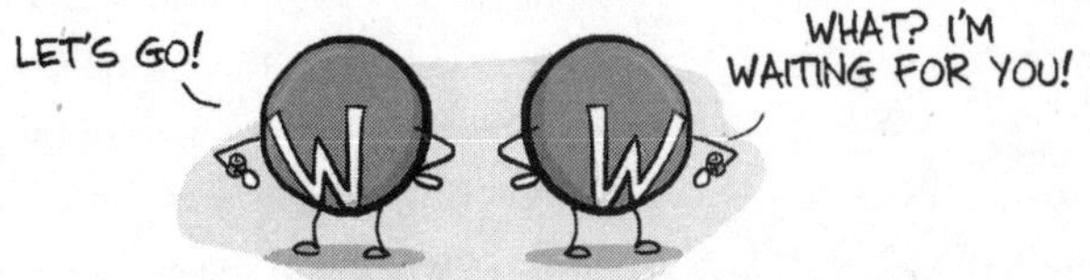

Question #4: Do We All Feel Time the Same Way?

Before the twentieth century, science considered time to be pretty universal: everyone and everything in the universe felt time the same way. It was assumed that if you put identical clocks in different parts of the universe they would continue to agree with each other forever. After all, that's what we experience in our day-to-day lives. Imagine the chaos that would ensue if everyone's clocks ran at different speeds!

But then Albert Einstein's theory of relativity changed everything by tying together space and time into one concept: space-time.[66] Einstein famously predicted that moving clocks run more slowly. If you take a trip to a nearby star by traveling close to the speed of light, you will experience *less time* than those left back on Earth. This doesn't mean that you feel time moving slowly, like in *The Matrix*. It means that people and clocks back on Earth will measure more time passing than the clocks on your spaceship. We all experience time the same way (at the normal one-second-per-second rate), but our clocks disagree if we are moving at high speeds relative to each other.

Somewhere in Switzerland, a watchmaker just had a heart attack.

Identical clocks ticking at different rates seem to defy all logic and reason, and yet this is what the universe does. We know it's true because we see it in our everyday lives. The GPS receiver on your phone (or your car or airplane) assumes that time moves slower for the GPS satellites orbiting the Earth (which are traveling at thousands of miles per hour in space curved by the gigantic mass of the Earth). Without this information, your GPS device would not be able to accurately synchronize and

66 Einstein's genius was not in naming things creatively.

triangulate your position from the signals transmitted by these satellites. The key is that while the universe follows logical rules, sometimes those rules are not what you expect. In this case, the culprit is the upper speed limit of the universe: the speed of light.

According to Einstein's theory of relativity, nothing, not even information or hot delivery pizza, can travel faster than the speed of light. This hard upper limit on speed (distance traveled per time) makes for some strange consequences that challenge our notion of what time is.

First, let's make sure we understand how this speed limit works. The most important rule is that this speed limit has to apply to anyone measuring any speed from any point of view. When we say that nothing can be observed to go faster than the speed of light, we mean *nothing*, no matter what perspective you have on it.

So let's do a simple thought experiment. Suppose you are sitting on your couch and you turn on a flashlight. To you, the light from that flashlight is zooming away from you at the speed of light.

But what if we strapped your couch to the top of a rocket and the rocket blasted away and started to move really fast? What happens now if you turn on your flashlight? If you point the flashlight toward the front of the rocket, does the light move at the speed of light *plus* the speed of the rocket?

We'll spend more time on these ideas in chapter 10, but the point is that in order for the light from that flashlight to appear to be moving at the speed of light to all observers (you in the rocket and the rest of us on Earth), something has to be different, and that something is time.

To make sense of this, it helps to go back to our thinking of time as the fourth dimension of space-time. And it helps to imagine that the speed limit of the universe applies to your total speed through both time and space. If you are sitting on your couch on Earth, you have no speed through space (relative to Earth), so your speed through time can be high.

But if you are on a rocket moving close to the speed of light relative to Earth, then your speed through space is very high. So in order for your total speed through space-time relative to Earth to stay within the speed limit of the universe, your speed through time has to decrease—as measured by clocks on Earth.

Still here?

It might bend your brain a bit to think that different people can report different passages of time, but such is the way of the universe. Even more bizarre, people can disagree in some cases about the *order in which things happened*, and all of them could be correctly reporting what they observe.

For example, two honest observers can disagree about who won a drag race if the observers are moving at very different speeds.

If you have a race between your pet llama and pet ferret, then, depending on how fast you are traveling and where you are relative to the race, you could see one or the other of your beloved pets win the race. Each of your pets will have their own version of events, and if your grandmother is capable of near-light-speed travel, she could disagree with all of them. And *they would all be correct!* (Note, though, that they would all also disagree about everyone's *starting* times.)

It's hard to swallow the idea that different people can experience time differently because we like to think that there's an absolute true history of the universe. We imagine that in principle someone could write down a single (very, very long and mostly superboring) story about everything that happened in the universe so far. If this existed, then everyone could check it against their experience, and barring honest mistakes and fuzzy vision, the story would agree with what people saw. But Einstein's relativity makes it clear that everything is relative, and even the description of events in the universe depends on who is recording that description.

Ultimately, we have to give up on the idea of time as an absolute single clock for the universe. Sometimes that leads us into areas that make no sense intuitively, but the amazing part is that this way of looking at time has been tested and shown to be true. As with many revolutions in physics, we are forced to divorce ourselves from our intuition and follow the mathematical path that is less influenced by our subjective experience of time.

Question #5: Will Time Ever Stop?

It's tempting to dismiss the notion of time stopping right out of the gate. We have never seen time do anything but go forward so how could it possibly do anything else? Since we have little idea why time is moving exclusively forward in the first place, it's difficult to say with confidence whether this will always be true.

Some physicists are convinced that the "arrow" of time is determined by the rule that entropy has to increase or that the direction of time is the same thing as the direction of increased entropy. But if that's true, what happens when the universe reaches maximum entropy? In such a universe, everything will be in equilibrium and no order can be created. Will time stop at that point or have no meaning? Some philosophers speculate that at this moment the arrow of time and the law of increasing entropy could reverse themselves, leading the universe to shrink back to a tiny singularity. But this is more in the category of late-night herb-inspired speculation than actual scientific prediction.

Other theories suggest that at the moment of the Big Bang, *two* universes were created, one with time flowing forward and one with time flowing backward. Even crazier are theories that propose more than one direction of time. Why not? We have three (or more) directions we can move in space—why not have two or more directions in time? The truth is that, as usual, we have no idea.

Time to Conclude

These questions about the nature of time are very deep, and the answers have the potential to shake the very foundations of modern physics. But while the scale of these questions makes them exciting to think about and ponder, it also makes them difficult to tackle.

How do you approach such a problem? Unlike other questions that we raise in this book, there is no clear experiment you can do to gain some answers. We can't stop time to study it, and we can't make repeated time measurements of the same event. This topic is so out there that very few scientists are working on it directly. It is mostly the province of emeritus professors and a few dedicated younger researchers willing to wade into such risky territory.

Perhaps we will make progress by tackling these problems head-on, or perhaps we will stumble upon a crucial insight when working on a different problem. Only time will tell.

9.

How Many Dimensions Are There?

In Which We Take Our Lack of Knowledge in New Directions

Gaining deep insights about the nature of the universe sometimes requires questioning basic assumptions and reexamining long-settled questions. For example, you might ask:

- Was JFK assassinated by aliens?
- Are there more dimensions to space than three?
- Is the universe powered by unicorns?
- Can you eat a pure marshmallow diet without gaining weight?

In most cases, the answer is "No" or "Please see a psychiatrist." But sometimes asking these questions cracks open a whole new way of thinking, one that can lead to new mind-blowing realizations that have a big impact on our daily lives.

If you are just now getting comfortable with the idea of space as a gooey physical thing rather than an empty backdrop to the universe, then take a firm grip on your mental safety railing and prepare to go even further as we explore the question of the *number of dimensions of space.*

Are there more dimensions to space than the three we are familiar with (up-down, left-right, and back-forth)? Could there be particles or beings that can move in these other dimensions? And if extra dimensions exist, what would they look like? Could we use them as storage for our shoes or to hide our extra stomach fat? Or build shortcuts to work or even to distant stars? These ideas sound absurd, but the truth of nature is no stranger to absurdity.

As usual, we have no idea what the answers are, but there are tantalizing theories that suggest extra dimensions might be real. So let's put our multidimensional glasses on and explore this potential hidden side (or sides) of our mysterious universe.

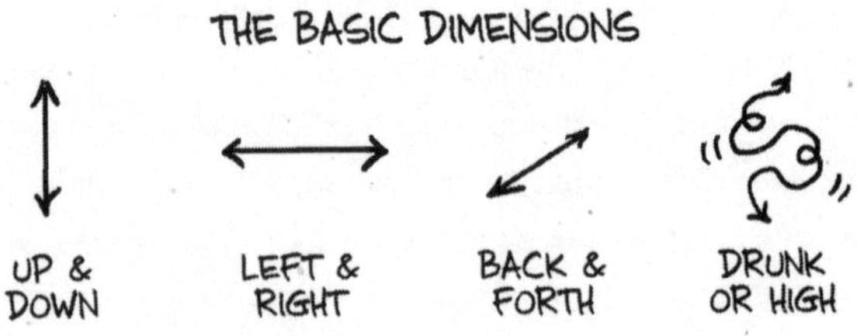

What Exactly Is a Dimension?

The first thing we should do is define exactly what we mean by a dimension. In popular books and movies, the word "dimension" is often used to mean a parallel universe: a disconnected existence where the rules are different and people can get supernatural powers or meet strangers who glow at night. Sometimes you can even open a "door to another dimension" to move between these universes. Those stories are a lot of fun, and parallel universes may yet exist, but scientifically the word "dimension" has a totally different meaning.

How can a word have one meaning in popular culture and another meaning in science? Most of the time, you can blame the scientists.

Whenever scientists need a word to describe a strange new thing they have discovered or imagined, they'll either (a) invent a new word (e.g., "exoplanets" to mean planets outside our solar system), (b) try to reuse a word that has a similar meaning (e.g., "quantum spin" to describe the physics of tiny particles that don't actually spin but do something that has similar mathematical properties as physical spin), or (c) borrow an existing word with a totally different meaning (e.g., the "charm quark," which is not very charming, or "colored particles," which have no color and seem politically incorrect anyway).

When you learn that the meaning of "dimension" in science doesn't mean an alternate universe where everything is made of chocolate and debts are paid in marshmallows, you might be tempted to wag your finger at those pesky scientists for stealing the word and giving it a different meaning. Well, put that finger away before you embarrass yourself because, in this case, the blame rests solely on the shoulders of science fiction writers. Mathematicians and scientists have been using this word with crisp precision for *centuries*.

In science and math, the word "dimension" means a possible direction of motion. If you draw a straight line, motion along that line is motion in one dimension.

In a world with one dimension, everything lives on an infinitely thin string. Because there's no other direction of motion, 1-D scientists could never cut in line or swap places with each other. They are like beads on a necklace or marshmallows on a skewer, always cursed to have the same pretty or sweet neighbors.

YOUR MANUSCRIPT IS SO THICK, YOU NEED ANOTHER DIMENSION TO PUBLISH IT!
OH, GO TO HILBERT SPACE!

HOW SCIENTISTS USE DIMENSIONS

Now draw a second line that is at right angles (ninety degrees) to the first line. We draw the second line at right angles so that motion along that second line is totally independent from motion along the first line. If the angle was less than a right angle, motion along the second line would also create motion along the first. These two lines create a plane, which lets you move in two dimensions.

So motion along a single line has one dimension, and motion on top of a flat plane defined by two lines has two dimensions. So far we have described a 1-D world (a string) and a 2-D world (a plane). To get a third dimension, all you have to do is draw another line that is perpendicular to the first two. In this case, it would be in the direction that points both above and below the flat plane.

This is what a dimension is: each one is a unique direction you can move around in so that motion in one direction is independent of motion in the other directions.

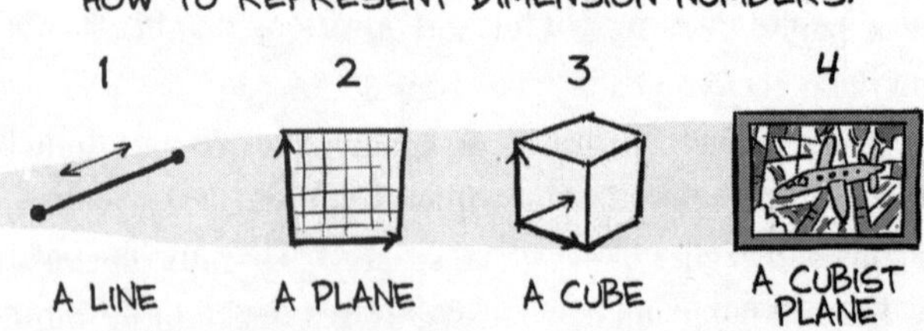

Can We Have More than Three Dimensions?

Drawing those three dimensions covers all the motion we are familiar with: up-down, left-right, and back-forth. There is no place in our 3-D world to fit a fourth perpendicular line, so our world seems to be very solidly three-dimensional, right? But physicists haven't come up with a good reason why we could *not* have more than three dimensions of space. In math, four dimensions would be just as good as seven or 2,035.

At this point, you might be thinking, *C'mon, if there were more than three dimensions of space, I would totally feel it!*

But would you? Could we tell if there were more dimensions? This is a question we can seriously ask. For example, what if the physical world has more dimensions but our minds are not capable of perceiving them? Although your mind is firmly convinced there are only three dimensions of space, it could be that we haven't noticed there are actually more.

Imagine if you were a 2-D physicist living on a flat plane, trapped the same way that all the words and drawings on this page are trapped on a flat sheet of paper. Your awareness and perception are limited to only what is on the plane (you can't "see" outside the page), so you wouldn't be able to tell if your flat world was actually floating inside of a 3-D world. In the same way, the 3-D world we know and love could actually be floating inside a higher dimensional space. This whole time, physicists in 4-D (or 5-D or 6-D) might be watching us and snickering at our limited perspective, the same way we might laugh at ants trapped in an ant farm.

But why wouldn't we be able to see or feel these other dimensions? That seems strange (and unfair) on the surface, but think for a moment about how your perception works. Our brain creates a three-dimensional model of the world inside our head because that is what has proven useful for survival on Earth. That doesn't mean that we are capable of perceiving the full nature of our environment. On the contrary, we are shockingly blind to features of our universe that may be irrelevant to daily survival but are crucial to understanding the fundamental nature of reality.

For example, you are very sensitive to light because it tells you a lot about where predators and marshmallows are. But you can't sense or notice the presence of dark matter, which surrounds you and holds important clues about how the universe works. Here is another example: you can't feel the 10^{11} neutrinos that pass through every square centimeter of your skin every second, yet if you could detect them, you might learn a lot about the Sun and particle interactions.

Every day we are bathed in information that is valuable for the modern physicist but that our bodies can't directly and naturally perceive. And this is because such knowledge is very difficult to collect or wasn't useful for survival on the marshmallow-strewn savanna of our evolutionary past.

I don't see any tigers.
OUR ANCESTORS
NOBODY'S ANCESTOR

So in response to the question "Can we have more than three dimensions?" the answer is yes. Mathematically speaking, there's no reason why there should be only three dimensions. It's possible that such a dimension could exist without our sensing it, especially if it's unlike any of our familiar three dimensions, but more on that in a moment.

How to Think in Four Dimensions

What *would* it be like to move through an extra dimension that is similar to our favorite three? It's difficult for us 3-D people to imagine what it would be like to move in anything but three dimensions. To help us grasp what that possibility would be like, let's take a step down in dimensionality and pretend that we are actually 2-D people who suddenly find ourselves moving around in a 3-D world.

If you were a 2-D person in a 3-D world, your 2-D body would still only be able to think and perceive in two-dimensional "slices" or planes within that 3-D world. Normally, that would be the limit of your

experience. But if you gained the power to move in the other dimension, the third dimension, you would now be able to float between different slices in that 3-D world. Your 2-D senses and mental worldview would not be able to sense your motion in that new direction, but if things were different in each slice, then you would perceive your 2-D slice-world changing around you. And if you could open your 2-D mind to a three-dimensional spatial concept (without inducing too many 2-D migraines), then you could stitch all those slices together to make a complete 3-D picture of the suddenly larger world.

Now use that idea and extrapolate it to our situation. If the world does have a fourth spatial dimension and we somehow gained the power to move through it, you could observe how the world changes along that direction of motion. Moving through that fourth dimension, you might see your 3-D world changing around you. If you have the brainpower and imagination, you could incorporate all of that information into a single holistic 4-D mental model.

In some sense, you sort of do this already. If you consider time to be a fourth dimension of motion, then the situation is very similar. The 3-D world around you changes with time, and in your brain you stitch together many different slices of time to form a four-dimensional (three spatial dimensions + one time dimension) picture of the world. You can't perceive all four dimensions simultaneously, but you organize 3-D snapshots along a timeline.

Where Are They?

You might reasonably ask: If there *is* a fourth spatial dimension (other than time), why do we never see it?

Well, we know that it has to be mostly irrelevant and useless to our survival in order to explain why we can't control or perceive our motion in that dimension. We also know that if it was a linear dimension like the other (regular) dimensions, we would probably have noticed by now. Even if we can perceive in only three dimensions, we would notice things appearing and disappearing if they move toward and away from us in this other dimension.

So we can be pretty confident that there's no fourth spatial dimension that is like the other three. If there is a fourth dimension, it has to be sneaky in some way that makes it hard for us to see it. One possibility is that all the force and matter particles we know about simply can't move through these extra dimensions of space. This would prevent objects from sliding in the fourth dimension and prevent energy (via force particles like photons) from dispersing into those additional dimensions. Could these impenetrable dimensions exist? Yes, but if they are truly impenetrable by any known particles, then we have little chance of discovering them or exploring them.

Another possibility is that these other dimensions are penetrable by only a select few particles, some of the rarer and harder-to-study ones, making them harder to notice. On top of that, these dimensions can hide in plain sight by being a little different.

How different? Imagine that these extra dimensions are actually curved and form little circles or loops. This means that motion through these dimensions doesn't get you very far. In fact, in a looped dimension, you end up coming back around to the same place you started.

If the idea of a curved dimension that forms a loop is strange nonsense to you, welcome to the club—it bends the minds of even the smartest

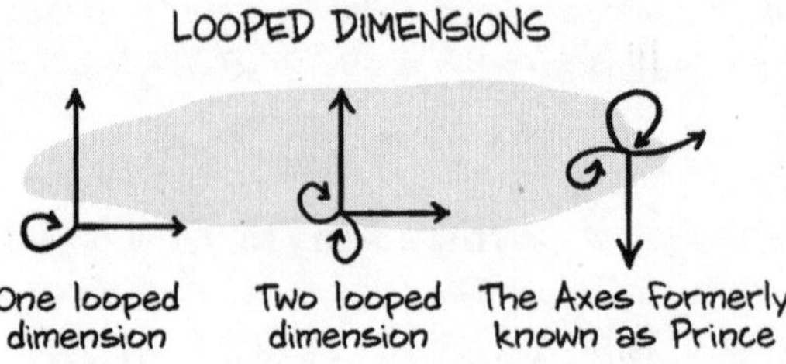

among us. In fact, it may even be possible that *all* spatial dimensions are actually loops. In the case of our familiar three dimensions, the loops would have to be very, very large—larger than the size of the observed universe (we discussed this possibility in detail when we talked about space).

If these extra dimensions are small and looped and only a few select particles can move in them, that would explain why we haven't noticed them. Things moving in these small looped dimensions wouldn't change very much in the three dimensions we *can* perceive, although there *are* ways to look for them, as we describe later in this chapter.

Do these extra dimensions exist? Are we actually in a universe that has more than three dimensions of space? The short answer is we have no idea. But there are actually good physics-based reasons why the universe might have more than three spatial dimensions. And even more exciting, we might have ways to discover them. Read on to find out how we might settle this question and still surprise those smug 4-D physicists who think we will never amount to anything.

Is It the Answer to Other Mysteries?

One of the biggest reasons why physicists believe there might be other dimensions is that their existence would help answer other deep questions we have about the universe. Namely, extra dimensions might explain why gravity is so weak.

If we compare the strength of gravity to the other forces, we find that gravity is more than just a little weak; it is *absurdly* weak. The other forces (the weak force, the strong force, and electromagnetism) have some differences between them, but compared to gravity, they are all muscle-bound bodybuilding superheroes while gravity is the Wonder Twins' pet monkey. Physicists really don't like to see this kind of disagreement. They are happy to disagree with one another about all sorts of things, but they expect harmony among the laws of physics. So one of the many questions about gravity is whether this unusual weakness is a clue that something else is going on.

Why is gravity so much weaker than electromagnetism and all the other forces? Well, extra dimensions might be the explanation. Most forces get weaker at larger distances. But just how quickly the strength of a force decreases with distance depends very specifically on the number of spatial dimensions there are. The more dimensions there are, the more a force can get *diluted* into all the different dimensions.

Think of what happens when someone farts at a party. If you are very close to the source, the smell is strong. But as you backpedal from the culprit, the stink molecules (i.e., fart particles, or "farticles") spread out into the air and get diluted.

Now, if the offending fart is released in a narrow hallway, everyone in

HE WHO DEALT IT, EXPERIMENTED.

that hallway will feel it strongly.[67] But if the fart is released at the intersection of several hallways, the fart will spread out in different directions and be felt less strongly by the people in those hallways. The rate of dilution depends on how quickly the volume of air grows, which gets bigger if there are more hallways.

HE WHO DEALT IT IN MULTIPLE DIMENSIONS DOES NOT SMELT IT.

Something similar happens with forces (without the smell). Suppose there are two extra spatial dimensions in addition to our existing three. Then the force you feel from an object (either gravity or electromagnetism) would spread into these other dimensions *in addition* to spreading in our regular three dimensions. As a result, the strength of the force would drop more rapidly as you move away from the source than you would expect if there were only three dimensions.

One caveat is that these extra dimensions have to be looped and small, less than about one centimeter, in order to explain why we haven't seen

67 In a one-dimensional world, there is no escaping farts.

them so far. And gravity has to be the only force that is affected by these extra dimensions, meaning that the other forces don't feel them.

So what happens if there are two extra loop dimensions one centimeter in size and only gravity can spread through those dimensions, not other forces? For objects less than one centimeter apart, the force of gravity would dilute into the extra dimensions and go down in strength very quickly. For objects greater than one centimeter apart, the extra dimensions wouldn't play a role. This would explain why gravity feels so weak to us: it is actually just as strong as the other forces for short distances, but once you go farther than one centimeter, most of it has already been diluted in the other dimensions.

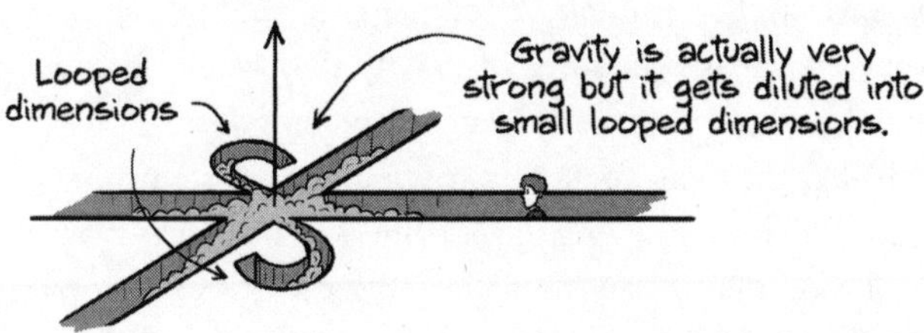
HOW EXTRA DIMENSIONS EXPLAIN THE WEAKNESS OF GRAVITY:
Looped dimensions
Gravity is actually very strong but it gets diluted into small looped dimensions.

Is gravity actually getting diluted like a fart in a hallway? We are not sure. The possibility of extra dimensions and their role in weakening gravity is still very much theoretical. Amazingly, however, we have ways to look for these extra dimensions.

Looking for New Dimensions

The idea that there might be extra dimensions sounds great because it would give a very simple and geometrical explanation for why gravity is weaker than the other forces. But right about now you should be thinking that it would be easy to check if this is correct: all you have to do is measure gravity at small distances, and if the force of gravity is stronger than expected, then surely that means those small loopy dimensions exist.

Unfortunately, it's not that simple. Measuring gravity may seem easy (after all, you measure it every time you get on a scale to weigh yourself), but that's only because we are used to measuring it at huge distances. When you step on a scale, you are measuring the force of gravity between you and the *entire* planet Earth, and one of you is huge.

EVERYONE'S FAVORITE GRAVITOMETER

Testing gravity at small distances, however, is a totally different animal. To test the strength of gravity between two objects one centimeter apart, you have to get their *centers* within one centimeter of each other, which means they have to be very small, so they can't have a lot of mass. And if the masses are small, the force of gravity will be so tiny it is almost impossible to measure (remember that gravity is weak). For example, if you put two ball bearings made of lead one centimeter apart, the force of gravity they would feel toward each other would be less than the weight of a speck of dust.

But here is the thing about physicists: if you say something is "almost impossible," that's only going to get them riled up. Add to that the possibility that such a measurement might prove the existence of extra dimensions, and you'll have a whole bunch of very smart people foaming at the mouth and coming up with mind-blowing measuring devices.

After much work in the past few years, physicists were able to measure how the force of gravity changes with distance at a scale of one millimeter. They found that, at least down to one-millimeter distances, the force of gravity still behaves just as it does for large scales. This doesn't mean that extra dimensions don't exist. It just means that if they exist, they are smaller than one millimeter in size.

Here's the other thing about physicists (there are many peculiar things about them; these are only two of them): until you make the actual measurement to confirm or deny a phenomenon, theorists are still free to speculate with abandon about how things might work. Physics can say things are true up to only the smallest possible precision we have in our experiments. Therefore, the only thing we can say with certainty at this point is that if extra dimensions exist that might be relevant to us they would have to be smaller than one millimeter in size.

POSSIBLE EXTRA DIMENSION
(ACTUAL SIZE)

Let's Blow Things Up

Measuring gravity is one way to check for extra dimensions, but it's actually not the only way. It turns out we can also look for extra dimensions using the power of particle colliders. Yes, those ten-billion-dollar twenty-seven-kilometer-long machines are good for more than just finding bosons named after Peter Higgs.

How can we use particle colliders to detect extra dimensions? Well, imagine that you have a tiny particle, like an electron, sitting in front of you. Maybe you have it resting in the palm of your hand. That particle is not only sitting there in our familiar three dimensions of space, it may also be moving *at the same time* along other extra dimensions. Remember that these other dimensions are loops, so the particle won't seem to be going anywhere in our dimensions, but it would be moving nonetheless. What effect would this extra motion have on our perception of this particle?

Well, if the particle is moving in the extra dimensions, it means it has momentum in these other dimensions, which means it has extra energy. But since the particle is not moving in *our* dimensions, we would experience that extra energy as extra mass (remember that mass and energy are the same according to Einstein). In other words, you would notice if a particle was moving in extra dimensions because it would be heavier than a particle that wasn't.

This is how we can use particle colliders to detect extra dimensions. If we smash particles together, and one day we see a particle that looks, for example, *exactly* like an electron (same charge, same spin, etc.) except it

is much heavier, we could reasonably suspect that it is actually an electron that is also moving in other dimensions.

In fact, if extra dimensions do exist, we could reasonably expect to find exact copies of all the particles we know about except they would be heavier due to their motion in extra dimensions. The theory predicts we would find "towers" (called Kaluza-Klein towers) of identical particles with heavier and heavier masses at regular intervals.[68] If we were to find such a sequence of heavier and heavier particles, it would be the smoking gun that confirms the existence of extra dimensions.

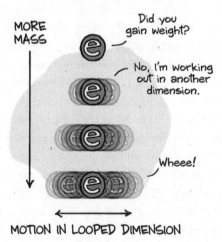

68 Muons and taus are not extra-dimensional versions of electrons, because they don't have a regular mass spacing and don't have the same weak-force interactions as electrons.

What Else Do Extra Dimensions Predict?

The existence of extra dimensions, even small looped ones, would have some other interesting consequences. If physicists are right that the weakness of gravity can be explained by its dilution into other dimensions, then that means gravity is just as strong as the other forces at small scales. Gravity may not be a weakling but rather a superstrong superhero *disguised* as a weakling.

This means that making a black hole might be easier than we previously thought!

Normally, you need an enormous amount of mass and energy in a small space to make a black hole. Particles, especially ones with the same electric charge (like protons), don't like being that close to one another. It takes a cataclysmic event (like a star collapse) to bring enough of them close enough together to reach the critical density needed to form a black hole. But if gravity is actually superstrong at small distances, then this extra gravity force could be strong enough to help protons form a black hole in simpler situations, say, for example, when you smash them together in a particle collider in Geneva.

So, yes, the Large Hadron Collider in Geneva could create black holes. If the scale of extra dimensions is about one millimeter, it is possible that the LHC makes one black hole per second.

But isn't that a terrible idea? Won't these black holes grow to gobble the Earth and all of our marshmallows? Relax, the answer is no. And if you have your doubts, there is a real website that you can check to see if the world has been destroyed.[69] Its makers promise to always keep it up-to-date.

Happily for our continued existence, the little black holes that the LHC might potentially be creating are different than the massive cosmic black holes made from collapsing stars. These are cute little black holes that will evaporate very quickly rather than gobble up Switzerland and the rest of the planet. Another reason you can relax is that very high-energy particles have been bombarding the Earth and colliding for eons, so if particle collisions were going to create planet-swallowing black holes, it would have happened already and we wouldn't be here.

IF YOU CAN READ THIS
THE WORLD
HASN'T ENDED

THIS BOOK DOUBLES AS A BLACK HOLE DETECTOR.

String Theory

Physicists are looking for ways to describe all of the fundamental forces (gravity, the strong force, the weak force, and electromagnetism) as part of a single comprehensive theory where everything is in harmony and no questions are left unanswered. Whether or not that is possible, it is a noble goal, and physicists have made considerable progress, though humanity is nowhere close to a final answer.

Along the way, though, some fun candidates have emerged. One of them is string theory, which suggests that the universe is not built out of zero-dimensional point particles but instead is constructed from tiny one-dimensional strings—not tiny like minimarshmallow tiny, but tiny like 10^{-35} meters tiny. The theory says that these strings can vibrate in lots of ways, and each vibrational mode corresponds to a different particle.

69 hasthelargehadroncolliderdestroyedtheworldyet.com

When you look at the strings from far enough away (a resolution of only 10^{-20} meters) they look like point particles because you can't see their true stringy nature.

One feature of this theory is that the math that describes it is much simpler and more natural if you have additional spatial dimensions. There are different flavors of string theory, and each predicts a different number of dimensions in our universe. Superstring theory prefers to work in a universe that has ten spatial dimensions. Bosonic string theory likes a universe with twenty-six dimensions. Where are these twenty-three additional dimensions and how did we miss them? This is like thinking your family had only four people and then finding twenty-two additional siblings hiding in the closet.

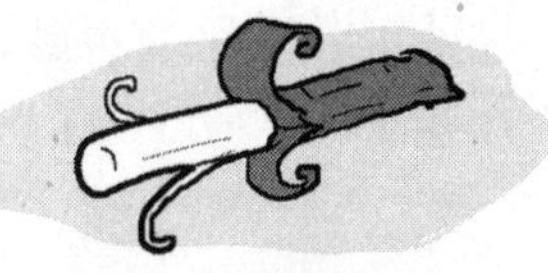
ALL CHEESE IS
STRING CHEESE.

WE'RE GONNA NEED
MORE CHEESE.

Like the theory that explains the weakness of gravity, string theories try to be consistent with our experience by making these new dimensions close up on themselves to form circles rather than making them infinitely long dimensions.

Wrapping Up These New Directions

Knowing how the universe's fundamental geometry is organized seems like a pretty basic part of understanding the world around us. There is an incomparable satisfaction that comes from discovering an unexpected truth about the universe and learning that the world we live in is different from the one we thought we did. Wouldn't you like to know if there was more to space than what you see and experience in your everyday life?

But finding extra dimensions could also have practical implications. We may discover that these extra dimensions are good for something. If they can store energy or give us access to regions of space we can't normally get to, who knows what we might be able to do with them.

IT'S AN INTER-DIMENSIONAL DE-FARTER.

Plus, discovering extra dimensions might give us clues that could help solve the puzzle of how the universe works (i.e., the other 95 percent of the whole universe). Even discovering that they *do not* exist would be significant. We could then ask why we have three dimensions (and not four or thirty-seven or a million). What's so special about three dimensions?

So far, the experiments measuring gravity at short distances have seen nothing unexpected, and the LHC has not discovered any black holes or particles moving in other dimensions. In other words, we have no evidence that this string-theory picture of the world is correct or that gravity moves in extra dimensions. So far, we really have no idea how many spatial dimensions there are in our universe.

Even stranger, it might be that the universe has different dimensionality in different regions—perhaps our little patch of space is 3-D but other parts of the universe have four or five spatial directions.

One thing is clear, though: the universe still has a lot of secrets waiting to be discovered. We just have to look for them in the right direction.

WHEEE!

10.
Can We Travel Faster Than Light?

No.

Okay, perhaps we should elaborate.

There are many things in physics that we are unsure about, but there is little doubt that nothing in the universe (light, spaceships, hamsters) can travel through space faster than the speed of light in a vacuum: 300 million meters per second.[70]

To put this in perspective, hamsters run at about half a meter per second (when they are in a hurry). The world's fastest man sprints at about 10 meters per second. The fastest speed by a person using a vehicle on land is 340 meters per second, and the space shuttle traveled at about 8,000 meters per second while in orbit, reaching about 0.0025 percent of the speed of light. You are not likely to come anywhere near this speed limit in your everyday life, but nevertheless it is there: an unbreakable rule—a constant reminder that even in this strange and wonderful universe there are limits.

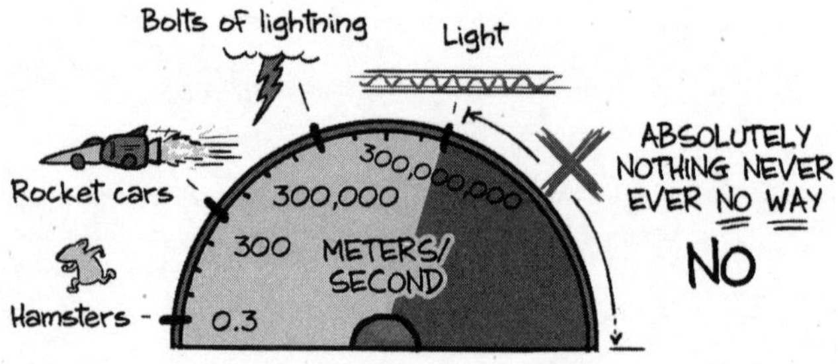

There is little doubt that this speed limit is real. The physics that describes it—relativity—has been tested repeatedly to very high precision. It is a basic principle woven into the fabric of modern physics theories. If this speed limit was not a fact of life, we would almost certainly have noticed by now. No matter what you do, who you know, or what you are, you cannot go faster than 300 million meters per second.

This maximum speed is a strange feature of our universe. As we will

70 "Through space" is an important caveat. Keep reading.

see, it leads to all sorts of weird consequences, from preventing different parts of the universe from ever interacting with one another to making honest people disagree about the order in which things happen.

And even though this speed limit is deeply enshrined in modern physics, there are still basic mysteries about it that puzzle physicists. For example: Why is there a maximum speed at all? Why is the speed limit 300 million meters per second and not 300 trillion, or 3 meters per second? Could the limit change? You better strap in, because we are going full speed through one of the biggest mysteries of the universe.

The Speed Limit of the Universe

When Einstein introduced the idea that there is a maximum speed in the universe, it was not very intuitive. After all, why *should* there be a speed limit in the universe? Why shouldn't you be able to hop on a rocket, blast off, press the accelerator pedal all the way down, and build up speed forever until you are zooming past galaxies left and right at ludicrous speed? If space is empty, what is actually preventing you from going as fast as you like?

This intuition that space is empty and that we can accelerate forever is where we get into trouble. As you might have learned in chapter 7, space is not an empty stage that you zip around in. Instead, we know that space is a physical thing, prone to bending and stretching and rippling, and it might actually take offense at your tearing across it at irresponsible speeds. In fact, it was learning about the speed limit of the universe that first gave physicists a clue that space was more than just emptiness.

So what do we know about this speed limit? First, it's not a hard stop. If you attempt to go faster than the speed of light, you don't hit some sudden hard wall or get pulled over by the galactic police. Your engine is not going to suddenly explode. Your Scottish engineer (whom you rudely call Scotty) will not start screaming at you that he doesn't know if the ship can take any more or not.

If you got on a spaceship and floored it, the following is what would happen: first of all, it would take you a *really* long time to get in the neighborhood of the speed of light. Even if you accelerate at 10g (ten times the force of gravity, or about 100 m/s^2), which is the maximum even top fighter pilots can briefly withstand, it would take you *months* just to get anywhere close to 300 million meters per second. And the whole time you'd be pressed against your seat—unable to scratch your nose or even go to the bathroom. It is not a pleasant way to go on a trip.

After you accelerated for a long time, this is what would happen: you would not go faster than the speed of light. That is basically it. Nothing dramatic happens; you just never get there. You would go faster and faster, but at some point, you would find that gaining more speed gets harder and harder. No matter how hard or how long you press the accelerator and/or how determined your facial expression is, you never reach or exceed 300 million meters per second. On the following page is a plot for the mathematically inclined.

What this chart says is that no matter how much energy you put into your engine system, your speed increases more slowly so you never quite reach the speed of light. It's like trying to go back to that svelte figure you

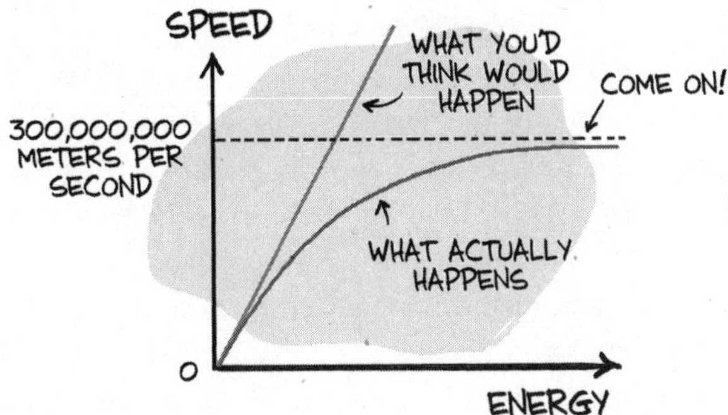

had in your twenties: it takes an impossible amount of time and energy, and you will never get there anyway.

It's *really* weird that the universe has a speed limit. Think about it: it means that when you try to move faster something prevents you from doing it even if there are no other forces being applied to you. It is an asymptotic limit built into the very fabric of space and time. In fact, it is happening right now as you walk down the hallway or drive in your car (hopefully listening to the audiobook, not reading while driving). As you may have noticed on the plot, the effect works even at lower speeds. It's not very obvious at low speeds, negligible even, but it is there. That means that relativity is not something that kicks in only when you're going close to the speed of light. It is always there messing with your motion, curving it just in case you ever do want to go faster than light. Think you can make that three-pointer? You better push the basketball just a little bit harder because space itself is trying to make you air ball.

The speed limit of the universe is not just an upper bound or a hard ceiling. It's a distortion of how velocities work in space compared to our intuition of how they should work. It is part of space and time, and it acts to limit all velocities in a strange way.

What's the Big Deal?

At this point, you might be thinking: *Well, fine, so we can't go faster than light. What's the big deal? I wasn't planning on going faster than sixty-five (okay, more like eighty-five) miles per hour anytime soon.*

That is true. The 300 million meters per second speed limit of the universe is not really going to affect you in your everyday life. But this speed limit has some profound implications for our view of the universe. Namely, we have to give up on the idea that time, and even the order that events happen, is the same for everyone everywhere.

Reasonable people can all expect that what happens happens and that we can all usually, given obvious evidence, agree on what happened. But that is not the case in this universe you were born in. Sequences of events can look totally different to different people, and it's all due to the speed limit of the universe.

To really understand how a speed limit in the universe can lead to such weird things happening with space and event planning, let's imagine a very common situation: suppose you give your pet hamster a flashlight. You know what, let's go nuts. Give your hamster *two* flashlights.

Now suppose your hamster points each flashlight to either side of her and turns them both on at the same time. Let's ask a very simple question: How fast are the photons from the flashlights going?

Easy, right? The answer is *c*: the speed of light (light is made of photons, remember?). Each photon is going in each direction at the speed of light. That is what your hamster would discover if she measured how fast

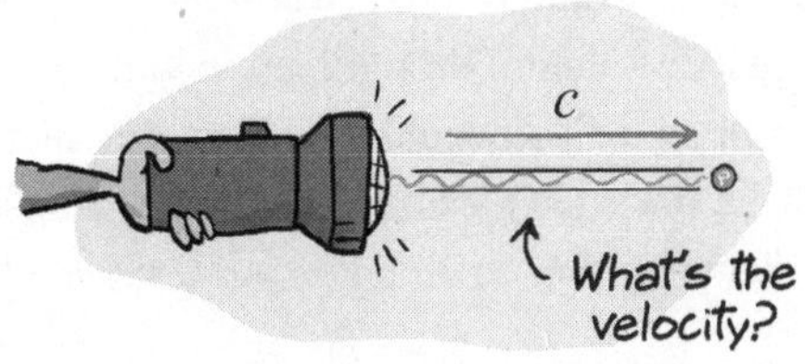

those photons were moving relative to the ground (we are, of course, assuming she has an advanced degree in experimental physics).

That makes sense, right? No controversy here. We can all agree that if you shine a flashlight (that shoots light) that light will go at the speed of light (hence the name).

Now take a mental leap and remind yourself that your hamster is actually standing on a giant ball of rock called the Earth that is hurtling through space. Then take a really big step back and imagine that you are out floating in space, wearing a spacesuit, and watching the Earth move past you to the right, carrying with it your beloved hamster and her two photon shooters (aka flashlights).

So you are watching the Earth move to the right with velocity V_{Earth}. Now let's ask: How fast do you (the astronaut reader) see those two photons moving?

If the photons are moving at the speed of light relative to Bertha (that's your hamster's name, by the way) and you are watching Bertha move past you, then your intuition would tell you to add the velocities together. So you might think that the photon on the right would have a velocity $c + V_{Earth}$ and the photon on the left would have a velocity of $c - V_{Earth}$. But if c is the speed of light, does that mean that you're going to see one photon move faster than light and the other photon move slower than light?

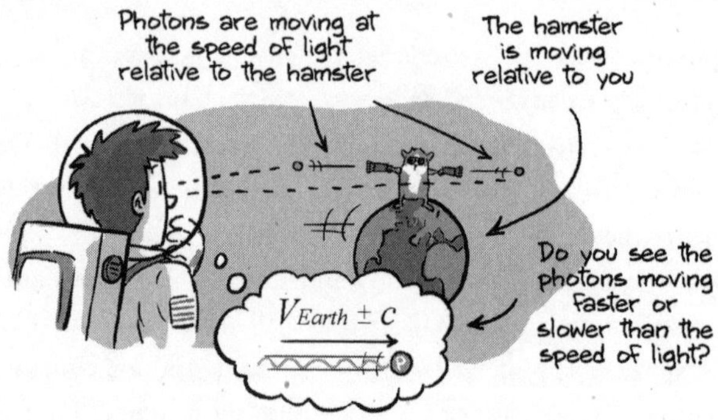

No! That's impossible, right? Nothing, not even light, can go faster than the speed of light (hence the name)! So what actually happens?

First, think about the photon that is moving in the same direction (to the right) as the Earth. This is the photon that your intuition would tell you should be moving faster than the speed of light. But because of the speed limit, you will actually see this photon traveling away at *exactly* the speed of light (relative to you). But this is weird because that's *also* what Bertha sees *relative to her*. Even though you and the hamster are going at different speeds, you both see that photon moving at the *same* speed relative to each of you.

How does that not break all logic and reason? What it really breaks is our expectation that everyone has to see things the same way. There is no getting around the fact that this makes for a weird universe with counter-intuitive phenomena.

Equally weird is what is happening with the photon moving to the left. You might naïvely expect that the photon would be moving slower than the speed of light ($c - V_{Earth}$) because the photon is coming from the Earth, which is moving to the right. But another weird property of massless

particles (like photons) in a vacuum is that they *always travel at the maximum speed allowed by the universe*. They never slow down.[71]

So light always travels at the speed of light regardless of who is measuring it and how fast they are going. That means that when you are floating in space seeing the Earth go by, you are going to see those two photons moving at exactly the speed of light with respect to you, and Professor Bertha on Earth is going to see those two photons moving at the speed of light relative to *her*.

That's one of the mind-blowing things about the speed limit of the universe: it applies to *relative* velocities between objects, not *absolute* velocities.

This is because there *is* no such thing as absolute velocity in this universe. You might think you are pretty special floating out there in space thinking you're an authority on how fast things are moving, but in actuality, you and the Earth are also moving relative to something else (say, the Sun or the center of the galaxy or the center of the cluster of galaxies we sit in). Even if there *were* a center of the universe (there isn't), who knows what your real velocity relative to *it* would be. So absolute velocities have no meaning.

The speed limit of the universe says that nothing can be *seen* to be moving faster than the speed of light. That's one of the weird things about it, and it's the reason things start to get even more bizarre.

71 What keeps massless particles (like photons) moving at the speed of light? As weird as light seems to be, it would be even weirder if it could slow down. If a massless particle could move less than the maximum speed, then a massive object could go fast enough to catch up to it. What would that look like? A massless particle is nothing but energy of motion (it has no mass). But if you could catch up to it so that it is not moving relative to you, then it has no motion and no mass, and so it is *nothing*. Poof. As weird as it is, it makes more sense for light to always travel at the maximum speed.

EVEN MORE
BIZARRE???

Things Get More Bizarre

Okay, so both you and your hamster see the light from the flashlights moving at the same speed even though the hamster is moving away from you. That's pretty weird, but it's about to get worse.

Suppose we put a target on either side of your hamster, and we ask the question: Which target will the photons from the flashlights hit first?

Not to scale.

If you ask Bertha, who sees the photons moving at the same speed in either direction, she will say that the photons hit both targets at the same time because both targets are the same distance away from her.

C
C
BOTH PHOTONS HIT THEIR
TARGETS AT THE SAME TIME

But that is different from what YOU see.

You see the two photons leave the flashlights at the speed of light (relative to you), but you also see Bertha (and the targets) moving. So while the photons are making their way to their targets, you will see one of the targets move *closer* to the photons, while the other target moves *away* from the photons. As a result, you'll see one of the photons (the left one) hit its target *before* the other photon reaches the other target.

In other words, you two see a totally different sequence of events! Bertha sees the light hitting the targets at the same time, but *you* see the light hitting one of the targets first. Here is the bizarre part: you are both right!

And it gets even stranger if you add more pets![72] Let's suppose that at the very moment you're discovering the bizarreness of the universe with your pet hamster, your pet cat (let's call him Larry) is returning home on his spaceship (the SS *Catnip*). He is returning in the same direction as the direction the Earth is moving relative to you (to the right), but he's currently moving *faster* than the Earth. So when Larry looks out the window of his ship, he sees Bertha and the Earth moving to the *left* relative to his ship.

72 This statement is always true.

Larry also sees Bertha's photons moving at the speed of light, as he must in order to maintain the speed limit of the universe, but since he sees Bertha moving to the left, he will report that the *right* photon hits its target first!

Now we have three conflicting reports: Bertha sees the light hit both targets at once; you see one of the targets get hit first; and Larry, who is probably surprised to find you out in space doing physics experiments, sees the other target get hit first. And you're all correct!

Not only do we have to accept that there is a top speed in the universe, but we have to give up the idea that events happen at the same time for everyone everywhere. No longer can we even assume the very reasonable-sounding idea that there is a single agreed-upon description of what happens in our universe. It all depends on which pet you ask!

HAMSTER DOUBLE-FLASHLIGHT EXPERIMENT SUMMARY

OBSERVER	ACTIVITY	OBSERVATION
YOU	CHILLING OUT IN SPACE.	THE PHOTON ON THE LEFT HITS THE TARGET FIRST.
BERTHA THE HAMSTER, PH.D.	WONDERING IF HER ADVANCED DEGREE IN PHYSICS IS BEING WASTED HERE.	BOTH PHOTONS HIT THEIR TARGETS AT THE SAME TIME.
LARRY THE SPACE-FARING CAT	COMING HOME TO RESTOCK ON YARN BALLS.	THE PHOTON ON THE RIGHT HITS THE TARGET FIRST.

History Is History

All of this should set off your crazy alarms right away. First of all, it means that there is no absolute ordering or history of events in the universe. Reasonable people (and their pets) can all correctly report a different accounting of what happened!

Think of it another way: you can change the order of events by watching them at different speeds. You, your hamster, and your cat all see the events happening in different orders because you are moving at different speeds. This is very counterintuitive because we like to imagine that the universe has a single common history: an ultimate time-ordered list of when things happened. But that is just not possible in our universe. The concept of a universal clock or universal simultaneity is gone—all as a consequence of having light travel at the same speed for everyone, which follows from having a maximum speed limit to the universe.

Breaking Causality

How far can you push this reordering of events? The fastest observer we have so far is the cat, and he sees the right photon hit its target first. What if the cat happens to be on a ship that actually *can* break the speed limit of the universe? As the cat goes faster and faster, he will start to see the time between the photon leaving the flashlight and hitting the target get shorter and shorter. At some point, Larry the cat will be going so fast that he actually sees the photon hit the target *before* it leaves the flashlight!

* THE PHOTON HITS THE TARGET BEFORE IT LEAVES THE FLASHLIGHT!?

But that makes no sense because that would violate causality (you know, the idea that effects are caused by causes and not the other way around). In a universe that doesn't have causality, things are crazy: water boils before you turn on the stove, pets lock you in the closet for neglect you are not yet guilty of. In such a bizarre universe, it is difficult to understand how things happen and it might be impossible to build reasonable laws of physics.

Incidentally, this is how we know that the speed limit of the universe is, well, universal. In 1887 two scientists named Michelson and Morley performed an experiment somewhat similar to our hypothetical hamster situation (albeit without the hamster). They shot a beam of light and split it into two perpendicular directions. Then they measured if the two resulting beams took the same amount of time to bounce off a mirror and return to their starting point. Like Bertha the hamster, they found that light took the same amount of time to travel in any direction. And because the Earth is moving at some unknown speed relative to the rest of the universe, they concluded that the speed of light is always the same no matter how fast or slow your relative motion is.

From this, we can conclude that *nothing* can go faster than light because it would result in situations where causality is broken (like Larry

We'll make "M&M" a household name!

THE MICHELSON-MORLEY EXPERIMENT

seeing the photon hit the target before it leaves the flashlight). And breaking causality is not a minor thing even for first-time offenders. The universe tends to take it pretty seriously.

Local Causes

So why is there a maximum speed? Why should the universe care how fast our cats and hamsters go? What possible purpose could this serve?

Can we derive this speed limit from any first principles or make sense of it in any way? The short answer is that we have no hard and fast reason why our universe has a speed limit, but there *is* a very good excuse to have one. A speed limit is useful in order to have a universe that is *local* and *causal*.

We talked about causality, and it seems like a reasonable requirement in a universe. By "locality," we mean the idea that the number of things that can affect you is limited to the number of things that are close to you. If there was no speed limit in the universe, then things that happen anywhere could have instant effects on the Earth. In such a universe, alien versions of the NSA could, in theory, read the texts (even snapchats) you send to your friends in real time, or alien scientists could develop tools that could instantly kill everyone on Earth. Instead, we have a rule that limits how fast anything (light, forces, gravity, selfies, alien death rays) can travel, which means that only things in your local environment can have causal connections to you.

If we want a universe in which we are not susceptible to instant weapons of mass destruction built by distant aliens, and in which cause and effect are respected, we have to accept some things that seem a bit odd, like people and pets disagreeing about the order of noncausal events.

But Why This Speed?

We argued that having *some* maximum speed makes sense in a universe that obeys cause and effect and locality.

But as is often the case with physics, answering one question leads to deeper and more basic questions: Why does the universe respect cause and effect? We can't expect that the universe was designed to be sensible to our particular minds.[73] Why do we have this particular maximum speed and not another?

The question of why the universe is causal is very difficult even to discuss, not to mention answer in a satisfactory way. Causality is built so deeply into our pattern of thinking that we can't just step outside of it and consider a universe without it. We can't use logic and reasoning to consider a universe without logic and where reasoning is impossible or inappropriate. This is certainly a deep mystery, and since science assumes causality and logic, it is possibly a question beyond the power of science to answer. It may be one that we never solve, or it could be tied inextricably to the thorny questions of consciousness.

73 Though you might reasonably argue that in a universe with cause and effect, intelligent life will discover it and build it into their systems of logic even if they don't understand where it comes from.

The more tractable question is: Why *this particular* maximum speed? None of our theories gives any reason for choosing one value over another. A causal universe with a faster speed of light would be less local than ours; a causal universe with a slower speed of light would be hyperlocal. But each of those universes would still work, and any setting of the speed of light is allowed in physics. It just so happens that we have measured it in our universe and found it to be 300 million meters per second: very fast compared to human experience but very slow compared to the distances one has to travel to move between the stars or the galaxies.

THE SPEED LIMIT OF THE UNIVERSE:
Just fast enough for us to see stars...
...but not fast enough for us to reach them.

Right now we have no idea why the speed limit is what it is. But we can speculate about different possibilities.

It could be that this is the only possible value and that the speed of light reveals something deep about the nature of the universe and space-time. For example, if space-time is actually quantized, then maybe the speed of light comes out of the way that information is transmitted between adjacent nodes of space-time. In the strings of a guitar, the speed of the waves along the string is determined by the thickness of the wire and the tension on the string. Something similar to that could be what determines the speed of light.

Or perhaps one day we'll come up with a unified theory of space-time that makes it obvious why light and information have to propagate at a certain speed, and all questions will be answered. But right now that seems about as likely as your pets preparing dinner for you.

On the other hand, it could be that the universe can have any value of the speed of light between (but not including) zero and infinity. Zero would correspond to a noninteracting universe and infinity to a nonlocal universe. If the universe could have had any value for the speed limit, how did this one get chosen? We seriously have no idea, and anyone who tells you that they do is either a time-traveling physicist from the future or has serious delusions of grandeur. Either way, do not ask them to watch your pets for you.

Maybe the speed of light is a local law of physics, not a universal one, which is valid in our part of the universe because of the way space-time congealed after the Big Bang ended. Perhaps in each region of the universe the speed of light is determined by random quantum mechanical processes. That would suggest that there are other parts of the universe with widely varying values of the speed of light. None of this even meets the standard of a complete idea, not to mention a testable scientific hypothesis. But it's fun to think about.

Past and Future

If we have no good reason why the speed of light is what it is, how do we know it will not change in the future or has not been different in the past?

We can't travel to the past to do experiments, but the universe has given us a beautiful gallery of ancient astronomical events: the night sky.

Remember that as we look out into the sky we are not looking at what is happening out there right now but at what happened in the past. The farther away an object is, the longer it takes its light to reach us and the older our image of it is now. We can effectively peer into the past by looking at objects farther and farther from us. Astronomers have applied our current laws of physics—including the speed of light—to the orbits and collisions and explosions that we see in the sky, and there is no hint of any violation of the universal speed limit.

When it comes to the future, predictions are difficult. We can extrapolate based on 14 billion years of history; that seems like a solid game to play, but it implicitly relies on the assumption that the universe will keep working the same way in the future that it has in the past. That is pure assumption—we know that the universe has had multiple radically different periods in its past (pre–Big Bang, Big Bang inflation, current era of expansion) so predicting that the universe will not change in the future smacks of overconfidence.

But Maybe We Can Visit Other Stars

Traveling faster than light is an intriguing possibility not because anyone wants to win a race against photons but because humans have a fundamental desire to explore the universe around us. Land on alien planets, visit distant suns, perhaps meet aliens and make friends with their silly pets—few people would turn down the opportunity to do any of these things.

Those of you eager to jump aboard the first spaceship to visit another star system or explore a neighboring galaxy will be sad to hear that a mere 300 million meters per second is the fastest we can travel in our universe. After all, the nearest star to our solar system is 40,000,000,000,000,000 meters away.

But maybe we are asking the wrong question. Instead of asking, "Can I travel faster than light?" what if we ask, "Can we travel to distant stars in a reasonable amount of time?" because the answer in this case is a very intriguing "Maybe, but it's very expensive."

Remember that the speed of light is the fastest that you (or me or your cat) can travel through space. But space is not an abstract backdrop of glowing yellow rulers. It's a dynamic physical thing with strange properties including the ability to expand and contract.

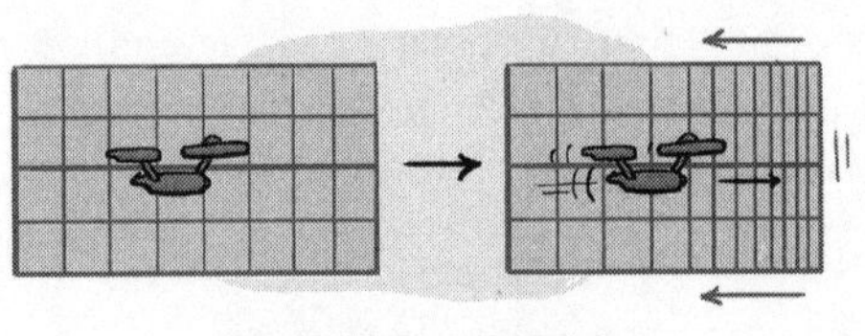

A WARPED IDEA

That last bit is crucial: What if we could *squeeze* the very space between us and some distant location so we get there in a reasonable amount of time without having to go very fast through space? Could that be done? *That* idea is a solid maybe. We have a lot to understand about the nature of space-time, but we know that it can be distorted and contracted. Unfortunately, doing so requires enormous amounts of energy, the equivalent of gazillions of hamster wheels spinning as fast as the hamsters' plump little bodies can manage. Scientists estimate that a warp engine that could compress the space in front of a ship would gobble unrealistic amounts of energy to get anywhere far away.

SHE'S GIVIN' IT ALL SHE'S GOT, CAP'N!

Or Maybe Wormholes?

Another way of shortening our travels without going faster than light is to use wormholes. Not the ones in the cute little worm farm you maintain to feed your pet lizards but the ones that are predicted by general relativity. Under the right circumstances, a wormhole in space could connect two places in the universe that are far apart from each other, allowing you to travel between them. In popular science fiction, traversing a wormhole involves crazy streaks of light, loud clunking noises, and an embarrassing loss of bladder control.[74] In reality, nobody knows what it would be like, and it might be nothing more than stepping through a doorway.

Indeed, if space has more dimensions than just three, it is possible that places that seem far apart in 3-D space are actually next to each other in other dimensions. Imagine if our universe was rolled up like toilet paper, with space looping around itself in layers. Things on the same sheet as we are on are what we typically think of as adjacent, but there could be other sheets nearby that could be traversed through wormholes that cut through the layers.

IT'S A CHARMIN' UNIVERSE.

Wormholes may sound like fantasy, but they're actually not inconsistent with any current law of physics. Unfortunately, all the calculations so far suggest that they would be very unstable, collapsing almost instantly, meaning that you'd hardly have time for an in-flight beverage before it collapses around you.

74 We made this up, but all the other bits about wormhole travel are made up, too, so why not.

In addition, we have no idea how to make wormholes, so we would have to trip over them and see where they lead. That is about as useful as stumbling around Manhattan blindfolded, getting into random strangers' cars, and hoping they are headed for Los Angeles.

Let's Keep the Dream Alive

Put aside the practical considerations—the impossible energy requirements and our lack of technology to create warp drives and wormholes—because these pesky details interfere with the awesome grandiose fantasy of interstellar travel that you, diligent reader, are entitled to after reading so many paragraphs that pour cold water on faster than light (FTL) travel.

The challenges of compressing space or traversing wormholes are immensely difficult, but take heart in the fact that physicists have upgraded the problem of interstellar travel from "totally impossible" to "very difficult and monstrously expensive," which is better than nothing.

Any prediction about the pace of technology far into the future would probably be either accidentally correct or embarrassingly naïve, so we decline to make any. But humanity's track record suggests that technological marvels are waiting for us in the future. And since there is no fundamental law of physics that prevents interstellar travel from becoming reality, there is still hope. When will it happen? We have no idea.

Muons Do It All the Time!

Physics is very careful about the fine print. Anytime there is a small loophole in one of nature's laws, you can bet there is a particle somewhere flouting it with abandon. Rereading the rules with a lawyer's eye, you might notice that the maximum speed limit is the speed of light *in a vacuum*. Why does it say "in a vacuum"? Because the speed of light depends on what it is passing through. The speed of light in air or in glass or in water or in chicken soup is less than the speed of light in a vacuum. The reason is that the photons have to spend time interacting with the pesky particles of chicken soup (let's call them "soupsons"), so their overall speed is slower.

So if you ask, "Is it possible to travel faster than the speed of light," then the answer is "Yes . . . technically." The technicality is that it is possible to travel faster than light does in some mediums—though still never faster than the speed of light in a vacuum. For example, a high-energy muon can pass through blocks of ice faster than light passes through ice. Technically, this is "faster than light" travel, though it seems lawyerly and unsatisfying.

FTL FTW

It won't help your dreams of starting your own colony on a distant planet and enshrining yourself as god of your own solar system, but it does make for some pretty snazzy effects. When a speedboat moves along the surface of a lake faster than the waves it makes in the water, those waves add up to make a wake. If an airplane travels faster than the speed of sound, it creates a shockwave of air called a sonic boom. What happens when a muon travels faster than light through a block of ice? It creates a light boom! This is also known as Cherenkov radiation, and the faint blue rings of light generated by this boom are routinely used by physicists to detect such particles and measure their speed.

So if the entire universe suddenly filled up with cosmic chicken soup (or ice), then *technically* it would be possible to travel through it faster than light and emit glowing blue rings all the way to your new home.

Summary

Can we travel faster than the speed of light?

Answer: yes, but no, but yes, but no.

11.

Who Is Shooting Superfast Particles at the Earth?

In Which You Learn That Space Is Full of Tiny Bullets

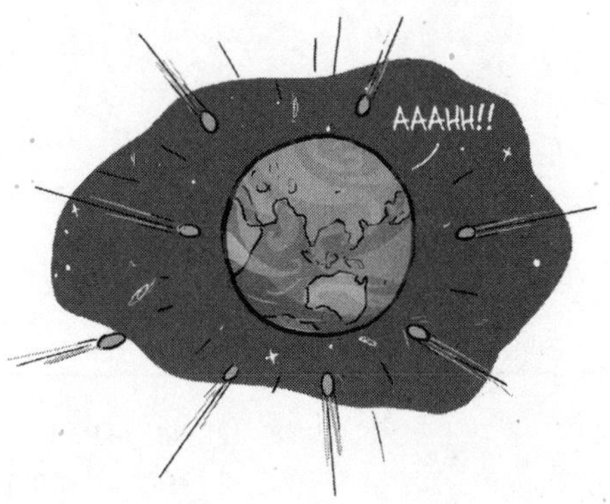

If you woke up one morning to discover that your house was being sprayed with bullets, that might qualify as an emergency situation. You wouldn't just relax, get dressed, and go about your day hoping that poorly funded scientists would eventually figure it out.

It just so happens that this is precisely your situation at this very moment—if you think of the Earth as your house and cosmic rays as bullets. Millions of these bullets hit our atmosphere every day, carrying with them more combined energy than an exploding nuclear bomb.

And the alarming thing is that we have *no idea* what (or who) is shooting them at us.

We don't know where they are coming from exactly, or why there are so many of them. And we don't know what process in nature could possibly be making such energetic ammunition. It might be aliens, or it might be something totally new that we've never seen before. The answer is beyond anything that even our overimaginative scientists can dream up right now.

So what are these enigmatic cosmic rays, and why are we being pelted with them at enormous energies? Find some cover and read on to learn more about this cosmic mystery.

What Are Cosmic Rays?

The name "cosmic ray" might be a little unnecessarily intriguing; it simply means a particle from space. Stars and other objects are constantly shooting out photons, protons, neutrinos, and even some heavy ions.

Our Sun, for example, is a major producer of space particles. Other than making the obvious visible light for which it became famous, the Sun also makes high-energy photons (UV light, gamma rays) that can penetrate far enough into your body to cause cancer. And that's nothing compared to the neutrinos that come from the Sun's fusion furnace: about 100 billion neutrinos from the Sun pass through your fingernail each second. Since neutrinos rarely interact with other bits of matter, this is not something you feel or have to worry about. Of those 100 billion neutrinos, only one of them on average will even notice that you are there and bounce off of a particle in your thumb. An average neutrino will pass right through the Earth without interacting, so while the bad news is that there is no shielding yourself from the zillions of neutrinos, the good news is that neutrinos really can't be bothered to hurt you.

Much more dangerous to the delicate machinery of human life are heavier, charged particles such as protons or atomic nuclei. A high-energy proton can tear through the human body causing significant destruction. Astronauts have to take special care and make sure that they are always shielded, which requires much more than slathering on sunscreen. Moreover, the Sun, like any enormous ball of fire, can be unpredictable. Most of the time it simmers nicely at a gazillion degrees, but sometimes it has indigestion that results in solar flares. These flares send strands of plasma far out into space and release extra doses of dangerous particles. Anyone spending time in space needs to have accurate predictions of the Sun's weather and get behind extra shielding very quickly when one of these flares is detected.

The point is that there are zillions of space particles hitting the Earth *all* the time. And they carry *a lot* of energy.

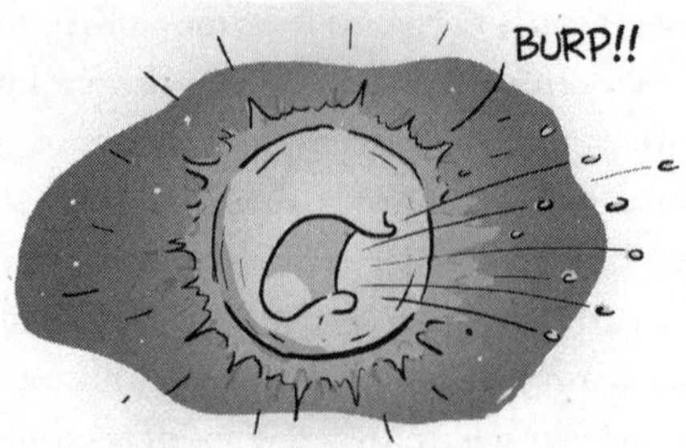

Fortunately for us on the surface,[75] we are mostly protected by the Earth's atmosphere. Most of the high-energy particles that hit the Earth slam into the air and gas molecules covering the surface of the Earth and break up, causing massive showers of lower-energy particles. If you ever wondered where the aurora borealis or aurora australis (i.e., the northern and southern lights) come from, they are the glow that comes from the stream of cosmic rays diverted by the Earth's magnetic field to the North and South poles.

But this protection only works when you are at the surface. If you spend any significant time far above the surface—as a flight attendant or a stowaway—you will receive more of this radiation. Unfortunately, wearing sunscreen on an airplane does not help.

How fast are these particles going? Down on the surface of the Earth, the world record for making fast particles is held by the Large Hadron Collider, which zooms them around at almost ten teraelectron volts (10^{13} eV). Anything with the prefix "tera" sounds impressive, but compared to the energy of the particles coming from space, it's pretty ho-hum. Cosmic rays hitting the Earth at the ten-teraelectron-volt energy level happen all the time. They are hitting the Earth's atmosphere at a rate of about one per square meter per second right now. If that sounds like a lot to you, it should, because they carry the energy equivalent of one slow-moving school bus raining down on every square meter of our world every second.

But then there are cosmic rays that hit the Earth at even *higher*

75 If you're reading this book on the International Space Station, *please* send us a picture.

energies—much, *much* higher energies. They make the particles we accelerate at the LHC seem like a baby crawling in slow motion through peanut butter. The highest-energy particle we have seen hit the Earth clocked in at over 10^{20} eV, which is almost *two million* times more energetic than the LHC's fastest particles. The record-setting space particle was going so fast that physicists nicknamed it the Oh-My-God particle. And when jaded physicists start sounding like flabbergasted teenagers, you know they are impressed.

Particles with this kind of crazy energy are surprisingly common. Almost 500 million of them hit the Earth every year. That's more than a million each day, or three hundred each second. Just now, while you read this sentence, over a thousand of them (the equivalent of the kinetic energy of two billion very-slow-moving buses) hit the Earth.

But here is the mind-blowing fact about particles this high on the energy spectrum: *we don't know anything in the universe that is capable of making such high-energy particles*.

That's right, we are being bombarded by millions of extremely high-energy particles on a daily basis, and we have no idea what could be creating them. If you ask astrophysicists[76] to estimate what the highest speed a particle *anywhere* in space could *ever* have (based on what we know right now), they will (a) thank you for asking them such a cool question,

76 We have.

(b) come up with crazy situations like particles surfing on exploding supernovas or black holes swinging particles around like slingshots, and (c) still come up short. Based on all the things we know about in the universe right now, the highest energy a particle could have in space is about 10^{17} eV, which is still more than a thousand times less energetic than the ones hitting the Earth every day.

Imagine if your Ferrari dealer told you that the car they sold you would max out at 200 miles per hour, and then you showed them that it could hit 200,000 miles per hour. You would conclude that even the world's Ferrari experts are more than a little clueless.[77]

That is the case with cosmic rays. There are cosmic rays hitting the Earth at energy levels that cannot be explained by anything we know in the universe, which can only mean one thing: there must be a *new kind of object* in the universe that we don't know about.

Okay, that seems like obvious logic when you write it down, but it's still a mind-boggling statement. Despite everything we know about the universe (at least 5 percent of it) and centuries of looking at the stars and building incredible high-precision tools, there are still things in the universe we haven't seen. Whatever is making these crazy-energy cosmic rays remains a mystery. And the fun thing is that the particles it sends us are clues about *where* the source is and *what* it could be, making this a specific puzzle that we can immediately sink our teeth into.

77 Yes, astrophysicists are Ferrari dealers in this analogy.

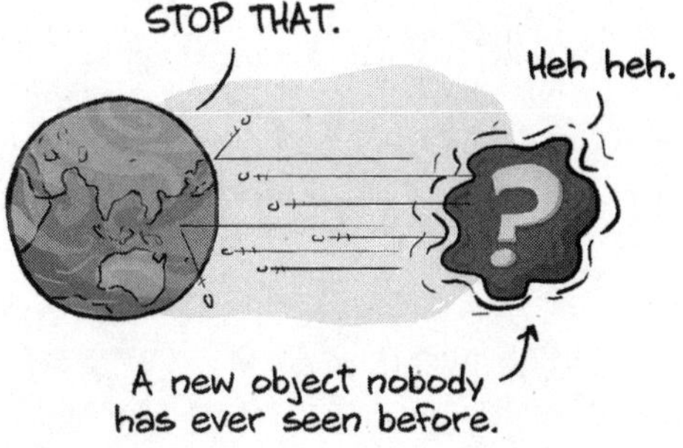

Where Are They Coming From?

If something was shooting super-high-energy anythings (snowballs, Fruity Pebbles, boogers, etc.) at you, the first thing you would do would be to look around and see where they were coming from. Are these crazy high-energy particles coming from a certain type of star? Or a supermassive black hole? Or perhaps an alien planet (or planets!)? Or maybe they are coming from *every direction.*

Luckily, the higher the energy of the particles, the more they will point back to what made them because very energetic particles will not be bent as much by any magnetic or gravitational fields between us and whatever is making them.

But to figure out where they are coming from, you need a few examples. It's like rooftop snipers; the more shots they fire, the easier they are to locate. The difficulty with nailing down where these cosmic rays are coming from is that the Earth is a pretty big target. Even though millions of them hit the Earth on a daily basis, actually placing a detector and catching them at the right time is tricky. We said earlier that hundreds of these hit the Earth every second, and we didn't lie, but the Earth is a very big place. So the more relevant number is how many cosmic rays hit an area the size of a typical detector, which is counted in square kilometers.

Particles at LHC energies (10^{13} eV) arrive at Earth at a rate of one thousand per square kilometer per second. Particles at absurd energies (10^{18} eV) arrive more rarely, at a rate of one per square kilometer per *year.*

But the prize jewels, particles above 10^{20} eV, are much rarer. They arrive at a rate of roughly one per square kilometer per *millennium.*

That makes it very difficult to figure out where they are coming from, because even if you build a very large detector, the chances that it will catch one of these high-energy particles is pretty slim. To date, we have detected only a handful of these superfast particles in all the cosmic-ray telescopes ever built. And so far, we can't pinpoint any source for these crazy space bullets.

The good news is that we do have an important clue about where they come from: they can't be coming from very far away. Visible light can travel billions of miles without scattering or being slowed down—that's why we can see distant galaxies despite their mind-blowing distance. Compare that to trying to see mountains just across the Los Angeles Basin and you'll be reminded how incredible it is that we can see so far through space.[78] But even though space seems very clear and empty to us, for an electrically charged high-energy particle it is actually like making your way through a crowded train station. The light that makes up the baby picture of the universe, the cosmic microwave background, fills the universe with a kind of photonic fog. Cosmic rays interact with this fog and get slowed down fairly quickly. A particle at 10^{21} eV can only go for a few million light-years before it gets slowed down to energies below 10^{19} eV or so.

78 You'll also be reminded that Los Angeles is not a great place to breathe the air.

This means that the high-energy particles we're seeing must be coming from a relatively nearby source, otherwise they would have been slowed down by the photonic fog. The only way they could have come from very far away is if they started out with absolutely *absurd* energies. If we can rule out the absolutely absurd, then we must conclude that whatever[79] is making them must be in our galactic neighborhood. That is a helpful clue because it removes a huge volume of space from contention, but it's also not that helpful because the volume of space that remains viable is still ginormous (scientifically speaking).

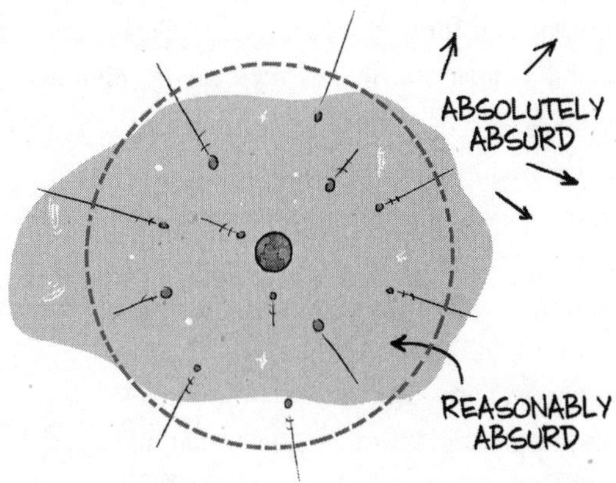

Altogether, these clues mean that we can be certain of this amazing statement:

79 Or *whoever* (dun-dun duuunnn).

SOME NEARBY OBJECT IS SHOOTING PARTICLES AT US WITH CRAZY HIGH ENERGY, AND WE HAVE NO IDEA WHAT IT IS.

That certainly counts as a cosmic clue that there are still new things to discover in the universe.

How Do We See Them?

When a super-high-energy particle hits the top of the atmosphere, it (thankfully) doesn't make it all the way down to the Earth's surface without banging into a lot of air and gas molecules. When a 10^{20} eV particle first hits a molecule in the atmosphere, it breaks up into two particles each with half that energy. Those two particles then hit other molecules, creating four particles with a quarter the energy, and so on. Eventually, you get trillions of particles with 10^{9} eV of energy washing over the surface of the Earth in a flash. This shower of particles is typically about a kilometer or two wide and consists mostly of high-energy photons (gamma rays),

electrons, positrons, and muons. Such a wide and powerful shower is how we know that a super-high-energy particle hit the Earth.

But seeing a mile-wide shower requires a very big telescope. Fortunately, while the telescope has to be very wide, it doesn't have to be continuous. Nobody can afford to build a mile-wide particle detector, so instead you take a chunk of land and dot it with smaller particle detectors. The Pierre Auger Observatory in South America is such a telescope. In 3,000 square kilometers of land, they have 1,600 particle detectors and more than 10,000 cows.[80]

WE'RE LOOKING
FOR MUUUUUUONS.

This detector is very good at seeing showers from ultra-high-energy cosmic rays, and it sounds very, very big because it is. But remember that in a square kilometer the super-high-energy particles arrive once every thousand years. So even if you cover 3,000 square kilometers, you might only see a few per year, which even after decades of observation may not be enough to solve the puzzle.

What else can we do? In order to narrow down the source and understand something about the origin of these particles, we will need a lot more examples. But building a bigger telescope using the existing technologies would cost a lot of money. The Auger telescope cost around $100 million.

One totally interesting idea is to try to find something that has already been built for other purposes and adapt it to work as a cosmic-ray telescope.[81] If you wrote down a description of the perfect cosmic-ray telescope, you would probably want it to have these features:

80 The cows serve no scientific purpose . . . that we know of.

81 Full disclosure: one of the authors of this book came up with this idea. No, not the cartoonist, the other one.

- Planet-wide coverage
- Rock-bottom price
- Totally awesome sound system
- Already built and deployed

Before you scoff at the absurdity of these specifications, consider for a moment whether this might be possible. Is there an existing network of particle detectors that are spread around the world and left unused for large portions of each day? If you just typed that question into Google on your smartphone, you are closer to the answer than you might imagine.

It turns out that the digital cameras in smartphones can work as particle detectors. The same technology that makes them great at taking pictures of your sushi lunch or your kids' latest incredible performance (really, your kids are amazing) also makes them sensitive to the particle showers produced when high-energy particles slam into the atmosphere. And smartphones are everywhere—there are more than three billion in active use as of this writing—*and* they are programmable, Internet connected, GPS enabled, and left unused all night long. If these smartphones ran an app that used the camera to detect particles, those phones could be part of a distributed, crowd-sourced, and global cosmic-ray telescope. Some scientists have recently proposed that if enough people (tens of millions) ran the app at night when their phones were not in use then the resulting network could see a lot more of those high-energy cosmic rays that we might otherwise be missing.[82] The more people who run the app, the larger the network and the more rays that can be collected. That could

82 And by "some" we mean Daniel and his friends. Visit the website http://crayfis.io for more info.

be you! You know you always wanted to be an astrophysicist, and if this crazy idea works, you could be part of solving one of the biggest mysteries in the universe.

What Could They Be?

When we say that astrophysicists can't explain the high energy of these particles, we mean that they can't explain it using only the objects we know about. If you give them free rein to invent *new* kinds of objects that might be making such speedy particles, then you get a lot of fun ideas.

Astrophysicists are creative people, and the history of our exploration of space has shown that the universe can be even more creative. Here are some ideas that could explain it—but remember that the most likely scenario is that none of these are correct and the real explanation is something even more mind-blowing than these crazy scientists could dream up.

Supermassive Black Holes

An explanation that was very popular for many years was that these high-energy particles were being created by incredibly powerful black holes at the centers of galaxies. These black holes have masses that are thousands or millions of times bigger than our Sun. Apart from the stuff that's already been sucked into the black hole,[83] there's a huge mass of gas

83 "Hole" seems like a terrible name for something that is actually very dense and solid. "Black mass" would be a better name if it didn't sound like a satanic ritual.

and dust swirling around it that's waiting in line to be sucked in. This stuff is under tremendous forces and has been observed to generate incredible radiation. However, the handful of very high-energy cosmic rays that we have seen in decades of observation don't seem to line up with the location of these active galactic nuclei. This means they are unlikely to be the explanation, clearing the way for even more outlandish ideas.

Alien Scientists

Some scientists wonder if we are the only intelligent species to be studying matter by trying to break it into little bits. What if aliens—yes, we mean intelligent extraterrestrial beings—have built a particle accelerator large enough to break down matter well beyond what we are capable of? The ultra-high-energy cosmic rays that we see could just be their leftovers, the pollution from their experiments. While we are on the topic of aliens, allow yourself to consider an even more fun and absurd possibility. What if the particles were found to be coming from a single location, such as a habitable planet surrounding a nearby star? What an amazing discovery that would be.

NOT COOL,
ALIENS.
NOT COOL.

The Matrix

And the ideas get crazier. Some scientists have speculated that our universe might exist solely as a simulation inside some cosmic computer. Beings in a larger metauniverse might be running some experiment using our universe.[84] How would we ever know? Such a simulation might have glitches due to the limitations of the computer that is running our universe.[85] If the simulation is done by chopping the universe into giant cubes and running a physics simulator inside each cube, then the simulation will give strange results for objects that move really fast across many cubes. In other words, patterns in the directions of ultra-high-energy cosmic rays could reveal our universe to be a simulation.

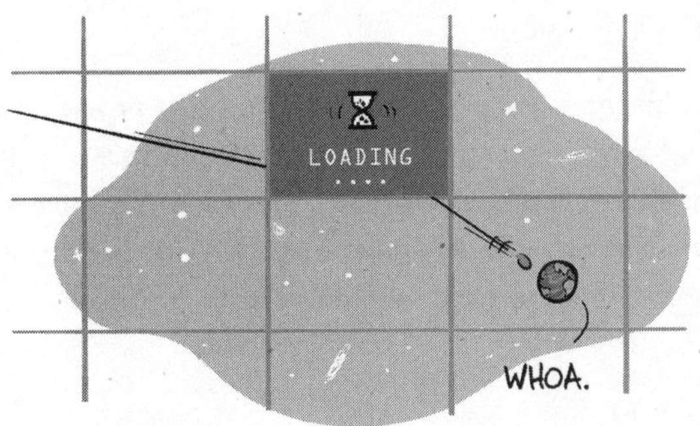

A New Force

We try to explain these particles using all of the cosmic objects and the forces in our physics toolbox. But the fact that they have remained unexplained for so long suggests another possibility, one that is both exciting and intriguing. Perhaps these particles are the result of some new undiscovered force. If such a force exists and is responsible for these cosmic rays, there would have to be a reason why we don't see its effects in other places. But the recent discovery that dark energy accounts for 68 percent of all energy shows us that it is not unrealistic to imagine that there are still universe-bending forces we haven't seen. Perhaps these particles are the clue that reveals to us an entirely new force of nature.

84 Yes, Douglas Adams had this idea first, but this is something taken seriously by serious scientists. Seriously.

85 If it's running Windows, let's hope it doesn't crash.

Regular Ol' Physics

It's possible, of course, that the answer is fairly prosaic and does not reveal a dramatic insight into the nature of the universe. It could be a new and currently unknown step in the life cycle of a star or some other object that is interesting to people who like studying stars, but it doesn't tell us anything deep about the universe. But let's keep the dream alive.

Cosmic Messengers

You have probably lived your whole life without knowing that you are being bombarded by superenergetic space bullets. If you hadn't read this chapter, you could have carried on and lived a happy life, blissfully unaware that there is something strange out there shooting at you and that nobody has any idea what, or who, it could be.

Well, it's too late for that. As you learned in chapter 8, you can't go back in time. But now that you do know, maybe you'll use this knowledge to look up at the sky a little more and be reminded that mind-blowing mysteries still abound in this universe.

Instead of thinking of these cosmic rays as bullets meant to harm you, you might think of them as messengers. Think about it: they travel through billions and billions of miles of space and bring with them information about some crazy new thing we have never before seen or even imagined. They carry proof of a process of stupendous energies and possibly also new forces, unknown cosmic mechanisms, or alien life-forms. They bring with them amazing discoveries.

And that is a bullet that you definitely don't want to dodge!

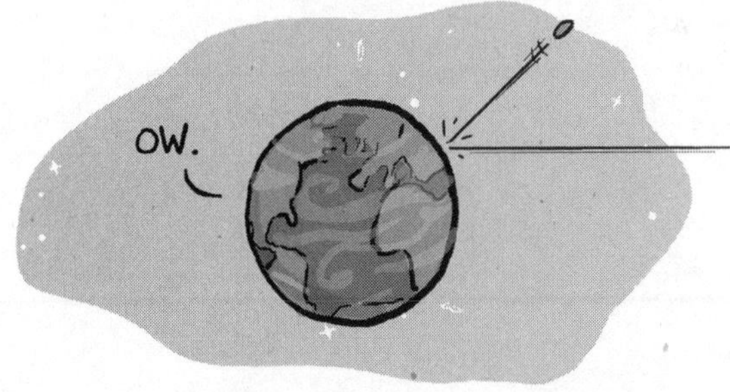

12.
Why Are We Made of Matter, Not Antimatter?

The Answer Will Not Be Anticlimactic

Math and physics have a very close relationship, which means that like longtime roommates they usually get along very well, but sometimes they fight about who ate whose leftovers.[86]

For example, physics depends on math to express the laws of physics, such as $E = mc^2$, and to perform important calculations, such as "what is the thickest slice of cake I can cut before my roommate notices?" Math is the language of physics the way English is the language of Shakespeare. If you don't know math, you would find reading a physics sonnet quite painful.[87] Actually, even if you do know math, poems written by physicists are not always very good.

On the other hand, math relies on physics to give it useful things to do. Without physics, math would be limited to abstract concepts, such as

86 Look, it's not physics's fault if math keeps leaving delicious chocolate cakes in the fridge for days.

87 "Shall I compare thee to an infinite sum of summer's days?"—from *The Lost Poems of Isaac Newton*.

imaginary numbers and large tax refunds. Physics can also excite mathematicians about discovering new kinds of math problems. For example, a lot of new insights in math have come from the development of string theory, a candidate for the ultimate physics theory.

There are also times when our intuition is an obstacle to understanding the physical world, in which case it is better to rely on math to guide us. For example, when trying to understand the bizarre behavior of quantum particles or income tax forms. In these situations, all you can do is follow where the math leads you. Assuming you crunched the numbers right, you can trust that math describes reality more accurately than your intuition. It might not make sense that you will get a twelve-quadrillion-dollar tax refund or that quantum particles can appear on the other side of impenetrable barriers, but if the math is correct, then that is what happens.

But not always. Sometimes math makes predictions that don't make physical sense and that we should reject. For example, let's say you run a cake company, and you are testing a new projectile delivery system for your chocolate cakes. How fast do you need to launch the cakes in order for them to follow a parabolic trajectory and land exactly on your customers' doorsteps? To calculate this, you would need to solve an equation that looks like this: $y = ax^2 + bx + c$ to figure out the shooting velocity and launching angle of your chocolate cake cannon. Because the equation has an x^2 in it, there are going to be two solutions for where the cake will hit the ground.

One solution will be the physical one, which will launch the chocolate cake in such a way as to perfectly deliver a devastatingly delicious dessert. The second solution, however, will give you a nonsensical answer: it will tell you that your initial velocity should be negative, which means you'd have to shoot the cake backward and directly at the ground. This is a correct *mathematical* solution, but not a physical one. It comes up because the mathematical approach uses a model of the problem that doesn't take into account all of the physical constraints of the system, such as the fact that cakes can't fly through solid ground. The whole thing also ignores the safety concerns of riddling the sky with chocolate cakes, but in this book, we care only about the physics.

In some cases, such as your soon-to-fail cake projectile idea, it is obvious that one solution is real and the negative solution should be ignored. Physicists have become used to this and routinely discard unphysical solutions as mathematical artifacts that are not real insights into our universe.

But be careful, smug physicists (and cake entrepreneurs), because some of those artifacts might be real, and Nobel Prizes (and profits) may be lying in wait. In this chapter, we will discuss how a negative solution led to the discovery of antiparticles and antimatter, and the questions about them that still linger today, nearly one hundred years after the last crumbs of the Nobel Prize–winning chocolate cake were illicitly consumed.

Mirror Particles

The whole business of antiparticles started when a physicist named Paul Dirac was working on equations to describe the quantum mechanics of electrons moving at very high speeds.

DIRAC FACTS:
• 1933 NOBEL PRIZE
• BECAME A PHYSICIST AFTER FAILING TO GET A JOB AS AN ENGINEER
• EINSTEIN THOUGHT HE WAS WEIRD

Earlier, physicists had found equations that could describe the quantum mechanics of lazy, slow-moving electrons. This was part of the mind-blowing quantum mechanics revolution of the early twentieth century, which required a complete rethinking of the nature of reality at the lowest levels. Quantum mechanics forced physicists to abandon deep and simple assumptions about the world: that things cannot be in two places at once or that precisely repeating the same experiment twice should give the same result. *Boom*. Mind blown.

EVERYTHING IS A WAVE.

But physicists in the early twentieth century were responsible for exploding our naïve perception of the universe not once but *twice*. On top of the philosophical craziness of quantum mechanics came the revolution that was relativity. Relativity shows us that the speed limit of the universe (see chapter 10) means that we have to abandon other long-cherished notions about the universe. In this case, the quaint idea that time is universal and that honest people will always agree about the order in which things happen.

Mind blown again.

Dirac took a look at these two crazy pieces of math—which correctly describe two different counterintuitive mind-blowing physics—and asked himself: What happens if I bring them together? If he was hoping for more craziness, he got what he was looking for.

He developed an equation (imaginatively called the Dirac equation) describing the behavior of fast-moving electrons that included both quantum mechanics and relativity; it was beautiful and elegant, and seemed to work except for one little problem.[88]

He noticed that his equations worked for the everyday negatively charged electrons, but they also worked for electrons with the *opposite electric charge*.[89] That is, his equation suggested that the laws of physics would work just as well for a positively charged electron, which he called the antielectron. This antielectron was just like the electron in many ways: it had the same mass and was described by the same quantum properties. But it had the opposite electric charge. This was puzzling because no such particle had ever been observed.

Some might have been tempted to disregard this as a mathematical artifact, a negative solution you should ignore. But Dirac was intrigued. What if this was more than math gone crazy, but something relevant to reality? After all, what physical law prohibited the existence of antielectrons? None that he was aware of.

In fact, Dirac looked at the equations and went even further. He proposed that *all* particles have a corresponding antiparticle.

So Dirac did more than predict one new particle; he predicted a whole

88 Note that he unified quantum mechanics with *special relativity*, meaning particles moving near the speed of light through flat space, not with *general relativity*, meaning particles moving in space distorted by large masses. That remains a puzzle.

89 Even more crazily, the equations also work for a normal negatively charged electron moving *backward in time*.

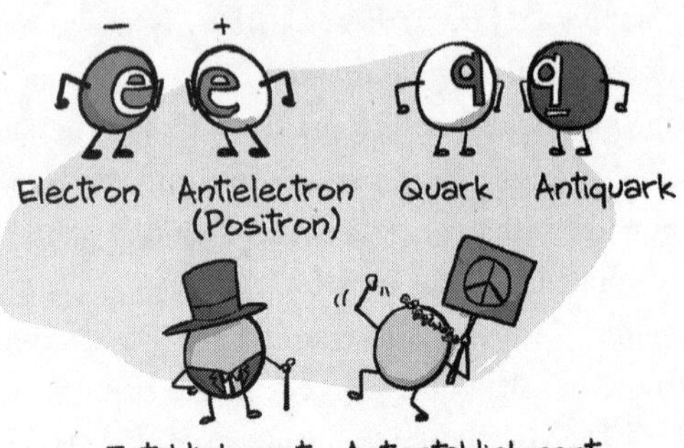

new *kind of particle*. That is no small idea. On the surface, it sounds crazy for every particle to have an opposite version of itself, like in a movie when a good character has an evil twin. In the case of particles, the antiparticle twin is not just different in electric charge but also in charges for the weak and strong nuclear forces. In a movie, this would mean that if the good twin is tall, fat, brunette, and likes dark chocolate then the evil twin would be short, thin, blond, and a fan of white chocolate (villainous!).

It's a crazy idea, but it also happens to be true. In fact, scientists have seen antiparticles many times. Shortly after Dirac proposed this idea, the antielectron (called the positron at this point) was detected. Today, nearly every charged particle we know of has been confirmed to have an antiparticle. Antiparticles can be easily produced in particle collisions, and at CERN a few picograms of antiparticles are produced annually. Cosmic rays from space also sometimes contain antiparticles or create short-lived antiparticles when they collide with the atmosphere.

Antiparticles are a good example of the symmetries that we find in physics at the smallest scale. You can think of each particle/antiparticle pair as two sides of the same coin instead of two unrelated particles. And remember that copies of particles happen in other ways in the organization of our universe: each of the matter particles already has two heavier cousins. For example, the electron has the muon and the tau particles, which have nearly identical quantum properties as the electron (same charges and spin) but have more mass. So the electron is copied in two ways: it has its heavy cousins and its antiparticle. And, of course, the heavy cousins have their *own* antiparticles.

It may not stop there! A speculative theory called supersymmetry proposes that every particle has yet *another* kind of mirror, a superparticle that is similar to the original particle (same charge and maybe the same mass) but has different quantum spin. The universe is full of funhouse mirrors that copy and distort the patterns in the particles in different ways.

PHYSICS: PUTTING THE "FUN" IN "FUNDAMENTAL."

But all these new particles only raise more questions: Why do we have these evil twin versions of our particles?[90] And why don't we see more of them flying around in our everyday lives?

Antiparticle Annihilation

Like anything that plays a prominent role in science fiction, some common misconceptions about antimatter can crop up. For example, you might have heard that when a particle touches its antiparticle they explode. That sounds ridiculous, doesn't it?

Actually, this one turns out to be true.

90 Besides the obvious TV ratings boost.

When a particle meets its antiparticle twin, they do more than just hug and get cozy: they *destroy each other completely*. The two particles disappear and their masses are completely converted into a high-energy force-carrying particle like a photon or a gluon. This is what we call "annihilation." All traces of the original particles are gone. This happens not just with electrons and positrons, but also when quarks meet with antiquarks, or muons meet with antimuons. Bring a particle and its evil twin together and expect a lot of drama and a big flash of energy. So the craziest-sounding feature of antiparticles in science fiction is actually true!

This is a really big deal because there is a lot of energy stored in mass. Albert Einstein famously established that mass and energy are related to each other by the equation $E = mc^2$. Note that in this equation the speed of light, c, which is already large at 300 million meters per second, is *squared*, so a little bit of mass carries a lot of energy. When two particles are totally annihilated, a huge amount of stored energy is released. To be specific, a *single gram* of antiparticles combined with a gram of normal particles would release more than forty kilotons of explosive force, which is more than twice as powerful as the atomic bombs dropped by the United States in World War II. A normal household raisin weighs about a gram—so a raisin plus antiraisin combination would be a dehydrated weapon of mass fruitation.

The concept of annihilation may seem strange to you since objects turning into blinding flashes of energy is not something you see every day.[91] So what does it mean for two things to annihilate? Do they get close

91 Except for fire, which is a chemical conversion of stored energy into light.

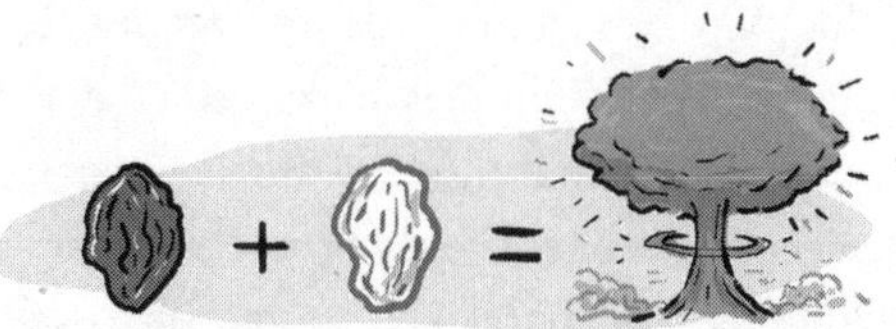

FRUIT CAN BE DANGEROUS

and then when they touch, *wham-o*, they suddenly turn into pure energy?

The first thing to keep in mind is that these particles are quantum mechanical objects, not actually tiny little balls. Sometimes you can use the tiny-balls picture to understand what particles are doing, and sometimes you have to use the quantum wave picture, but both are awkward and occasionally inappropriate. Like that one uncle at the annual family picnic. You know which one.

When two particles get close enough to each other, they don't actually touch, because they don't actually have surfaces. Instead, you can think of their quantum mechanical features as merging and the two particles as disappearing into another form of energy, in most cases a photon. From this energy, other kinds of particles can emerge, depending on the amount of energy you smooshed together. This is exactly what happens when we smash particles at the Large Hadron Collider to create new kinds of particles from ordinary everyday particles.

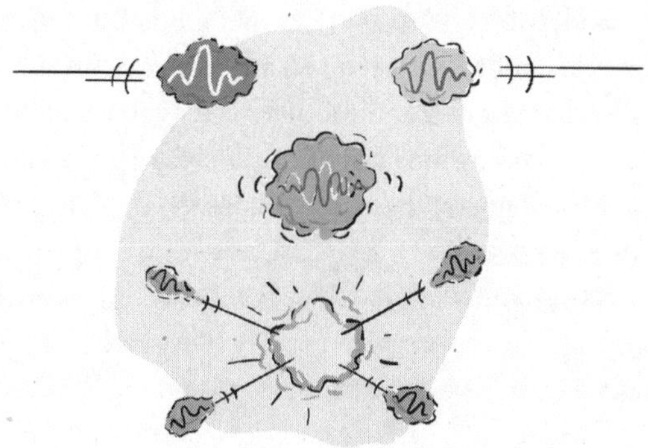

PARTICLE SMOOSHING

This means that, in a way, all particle interactions result in annihilation of the original particles into new particles. What's different about particles and antiparticles is that they are mirror versions of each other, which means that they have opposite charges. This makes them attractive to each other, so that they are more likely to smoosh together. At the same time, they perfectly complement each other, which means they can annihilate into something neutral, like a photon.

The other thing to keep in mind is that when particles interact (or smoosh, as it were), certain things are *conserved*. For example, we have observed that electric charges are never created out of nothing and that they are never destroyed. The total electric charge of the particles before and after the smooshing has to be the same. Why is that? We don't know. We don't understand why these rules apply; we simply see these patterns in experiments and incorporate the rules into our theories.

When an electron and its antiparticle, the positron, get close to each other, their opposite charges (–1 and +1) pull them in even closer. And once they smoosh, their opposite electric charges perfectly cancel each other out, allowing all traces of their existence to disappear so that only photons come out at the end. If you tried to do this with any other particles, say two electrons, their negative charges would repel each other. If you somehow managed to overcome their repulsion, there would be a net negative charge (–2) that would need to be conserved after the smooshing, which wouldn't allow total annihilation into a neutral photon.

And electric charge is not the only thing we have seen be conserved. You might wonder if any two particles with equal but opposite charges can annihilate each other (for example, an electron with charge –1 and an antimuon with charge +1). But the answer is that you cannot. There seems to be another rule in our universe about smooshing that says that "electronness" and "muonness" have to be conserved. You can't destroy an electron with a nonelectron. It only works with its antiparticle, the positron.[92] The same goes for all the other cousins of the electron: the muon and the tau.

92 Or with the electron neutrino, which also has electronness. An electron plus anti-electron neutrino can make a W boson.

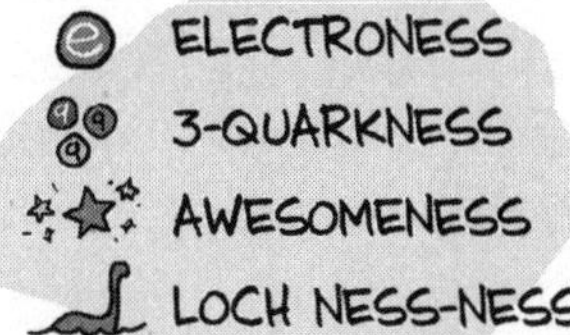

QUANTITIES CONSERVED DURING ANNIHILATION:
ELECTRONESS
3-QUARKNESS
AWESOMENESS
LOCH NESS-NESS

And it doesn't stop there. There is a whole list of conserved quantities (such as preserving the number of particles made of three quarks, or "three-quarkness"), each of which comes from observations about which particle interactions happen and which do not.[93] These rules seem to limit total annihilation to only particle/antiparticle smooshing.

Why does the universe have these weird rules? We don't know. Maybe one day we'll be able to show that these rules are a natural consequence of some underlying simpler theory of particles. But for now, it certainly suggests that antiparticles hold some important clues about the basic rules of the universe.

An Anti-You

So antiparticles are the strange shadowy twins of particles, and together they annihilate each other like tiny mixed-martial-arts fighters dueling to the death. Believe it or not, it gets more interesting than that.

It turns out that antiparticles can assemble themselves just like regular particles can to make antiversions of more complex particles like neutrons and protons. For example, you can make an antineutron by combining two anti–down quarks and one anti–up quark. The resulting antineutron is still electrically neutral (like the neutron), but its insides are made of antiparticles. And you can make an antiproton by combining two anti–up

93 Particles with three quarks (like protons and neutrons) are called baryons, so "three-quarkness" is usually referred to as "baryon number."

quarks and one anti–down quark. An antiproton is like a proton except it has negative charge because its insides are also made from antiparticles.

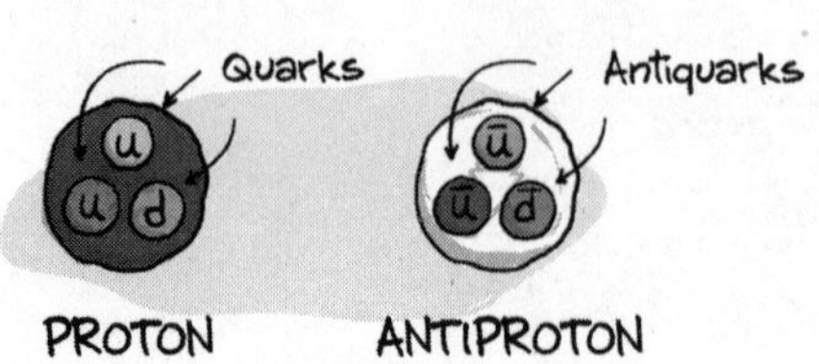

And it gets even weirder. Once you have antielectrons, antiprotons, and antineutrons, you could potentially make antiatoms! A positive electron and a negative proton would behave just like their regular counterparts except with the charges reversed. If you get an antielectron together with an antiproton, the antielectron would orbit around the antiproton and you would get antihydrogen!

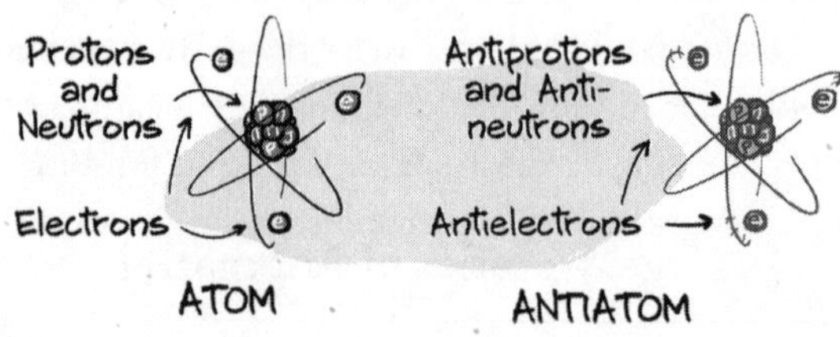

In theory, if you assemble enough antiparticles together, you can make antianything. For example, perhaps you can combine two antihydrogens with one antioxygen to get anti-H_2O or *antiwater*. Antiwater would look and feel and behave the same way as regular water except that if you drank it you would explode in a blinding flash of light, which, we admit, would be antirefreshing.

But why stop there? If you could make antiwater, you could potentially also make antiversions of any atom and any molecules. Perhaps even antichemistry and antiproteins and anti-*DNA*.

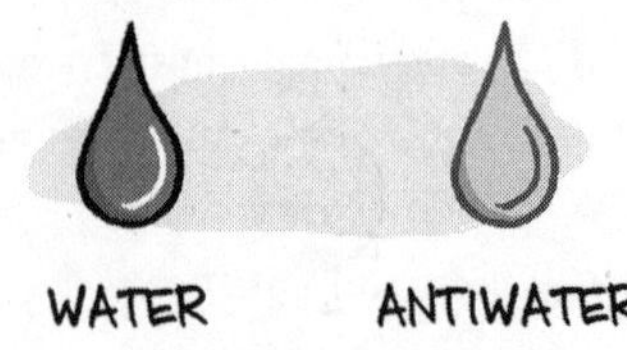

There could be a whole other Earth or a whole other you that looks exactly like you except it's made of antimatter. Antishe or antihe could be driving an anticar and live in an antihouse and even be reading an antiversion of this book that is made out of antipaper and is filled with jokes that are actually funny.[94]

In fact, there is nothing fundamentally "mattery" about our kind of matter and there is nothing "antimattery" about antimatter. If the situation was reversed and we were somehow made of what we call antiparticles, then we probably would be calling the antiparticles "matter" and the regular ones "antimatter"—since those are just arbitrary names. In other words, *we* could be the evil twins! (Cue shocking reveal music.) Wouldn't that be the ultimate twist ending?

Of course, all this talk of antiparticles and antimatter only begs the question: Where is all this antimatter?

The Mysteries of Antimatter

We know that antiparticles exist, and Dirac's formula does a great job of describing their behavior at high speeds. But that doesn't mean we fully understand them. In fact, this strange phenomenon of our universe raises more questions than it answers.

94 And with antifootnotes, negatively numbered.

For example: Why do antiparticles exist? Our modern theory of particles requires them, but you could also imagine other theories that include more kinds of weird twins (evil triplets or nefarious quadruplets, perhaps).

Other questions include: Are antiparticles exactly the opposite of regular particles or are there subtle differences in behavior, texture, flavor, or chocolate preference? Do antiparticles feel gravity the same way as particles, or do they feel it the *opposite* way?

But the biggest of these questions is a simple one: Why is our world made of matter, not antimatter?

If you are positive you can handle some negativity, then read on to learn more about these mysteries. It's . . . free of charge.

Why the Universe, Not the Antiuniverse?

There *is* one *very* big, very important, and very obvious difference between matter and antimatter: matter is everywhere and antimatter is almost nowhere to be found. That is, the universe seems to have a lot more matter than antimatter.

If matter and antimatter are equal but opposite versions of each other, then we would expect that the same number of particles and antiparticles would have been created during the Big Bang. But play that scenario out for a moment and see where it leads you: if there was an antiparticle created for every regular particle, then eventually all the particles would meet with their antiparticles and annihilate each other, converting all matter in the universe into photons. Since you are alive and reading this book, and you're pretty sure you're not made out of light,[95] we know that this didn't happen. Therefore, there must be some preference for matter over antimatter.

There are (at least) two possibilities to explain this inequality:

95 You're awesome, but you're not *that* awesome.

Possibility #1

During the Big Bang, slightly more matter was created than antimatter. And while the vast majority of the matter and antimatter annihilated itself into oblivion, the tiny bits of matter that were left over when the antimatter was used up is what was left to create all the galaxies, stars, chocolate cakes, and dark matter that exist today.

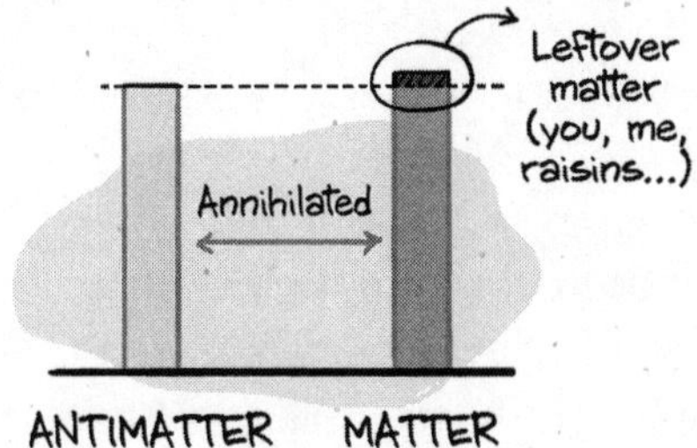

This possibility explains what we see, but it punts on the core concept. It turns the question "Why is the universe today made of matter, not antimatter?" into the equivalent question "Why did the universe begin with more matter than antimatter?" Unfortunately, we have no idea how to answer this question either. (In addition, most modern theories of the early universe are inconsistent with any asymmetry in the initial production of matter and antimatter.)

Possibility #2

During the Big Bang, the same amount of matter and antimatter was created, but over time something about the particles themselves caused there to be more matter than antimatter.

This is possible if there are physical reactions that destroy antimatter faster than matter or create more matter than antimatter. Since particles are created and destroyed all the time, even a very small difference in the

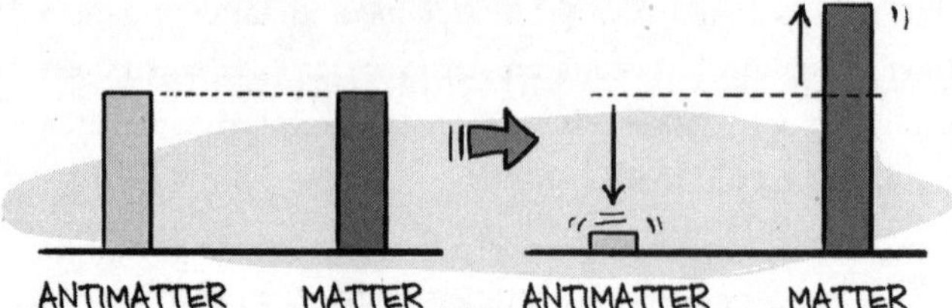
POSSIBILITY #2: THERE WERE EQUAL AMOUNTS BUT OVER TIME ALL THE ANTIMATTER DISAPPEARED
ANTIMATTER
MATTER
ANTIMATTER
MATTER

way particles and antiparticles are created or destroyed could add up to a huge imbalance.[96]

So possibility #2 seems promising. But how likely is it that the universe has an inherent preference for making or preserving matter rather than antimatter?[97] Most of physics is totally symmetric. And as far as we can tell, anything regular particles can do, antiparticles can antido. For example, a neutron can decay into a proton, an electron, and an antineutrino (this is called nuclear beta decay and it happens all the time). In just the same way, an antineutron can decay into an antiproton, an antielectron, and a neutrino.

Perhaps this preference is very small. In studying the creation and destruction of particles, physicists look for little imbalances between particles that oscillate between matter and antimatter versions of themselves. Unfortunately, while there is evidence for some inequality, it doesn't come close to accounting for the huge imbalance that we see today.

HAVE YOU TRIED BEING MORE POSITIVE?

So there must be something else going on that can explain the preference for matter over antimatter. Whatever it is might also give us a

96 The universe takes no vacations. Ever.

97 And if you think that an asymmetry in matter/antimatter creation and destruction is just as strange as an initial asymmetry in the amount of matter and antimatter created during the Big Bang, you have a fair point. But in the former case we would be able to test it and study it today.

clue as to why there are two classes of particles in the first place. But so far, we have no idea what it is.

Wait, Maybe Antimatter Is Somewhere Else

Maybe we have it all wrong. What if there *are* equal amounts of matter and antimatter in the universe, but they're all separated into different regions? The Earth and its immediate neighborhood are definitely made of matter, but what if there are *other* neighborhoods out there made of antimatter?

Matter and antimatter are so similar that we can't tell if a distant star is made of matter or antimatter just by looking at the light that comes from it. Both types of star would have the same nuclear reactions and generate photons the same way with the same energies.

So let's start closer to home. We know there are no significant quantities of antimatter on Earth because the Earth is made of matter and any antimatter on it would react explosively. Let's take a step further: Could there be big regions of antimatter in the space near the Earth? Could one of the planets in our solar system be made of antimatter?

Definitely not! Remember what happens when matter and antimatter get together: it is more explosive than political conversations with relatives. For example, if the moon was made of antimatter, then every time it was hit by a matter meteor there would be an enormous explosion and a giant flash of light. A meteor the size of a raisin would cause an explosion as dramatic as an atomic detonation. And the Earth and moon are constantly bombarded with matter meteors, small and large, so we know at least the moon isn't made out of anticheese.

The same argument goes for Mars and the other planets in our solar system. If Mars was made of antimatter, we would see the exploding photons all the time. In fact, if there was any significant concentration of antimatter near a region with matter, you would see constant annihilations and releases of photons at the border between the matter and antimatter regions. We see nothing like this in our neighborhood, so we are confident that our solar system is made of matter.

And remember, we've also sent matter-based objects (including people) out to explore our solar system, and none of those have been instantly annihilated in brilliant flashes of light.[98]

THE FACT WE DON'T SEE GIANT EXPLOSIONS EVERYWHERE MEANS THERE AREN'T BIG REGIONS OF ANTIMATTER OUT THERE.

Astronomers have expanded this search, looking for entire solar systems made of antimatter in our galaxy. So far, we have not seen the bright flashes of photons you would expect to find at the interface between the matter and antimatter regions. They have even considered the possibility of entire galaxies made of antimatter. But if any existed, we would see the space between the matter galaxies and the antimatter galaxies light up from the annihilation of particles streaming from both types of galaxies.

98 Yet!

Currently, astronomers have pushed this technique far enough that they are confident that our entire cluster of galaxies is all made of matter.

So far, that's the limit of our direct observation. Beyond that, we can't say for sure because the voids between clusters of galaxies are large enough that if there was a boundary between matter and antimatter out there it would be too faint to see.

Despite this, it seems likely that the rest of the universe is also made of normal matter. A universe organized into clusters of matter galaxies and antimatter galaxies would have required the matter and antimatter in the early universe to be widely separated, which would raise a whole new set of questions.

To recap, we have no evidence for large clumps of antimatter anywhere in our observable universe. So the question of why we see only matter and not antimatter remains open.

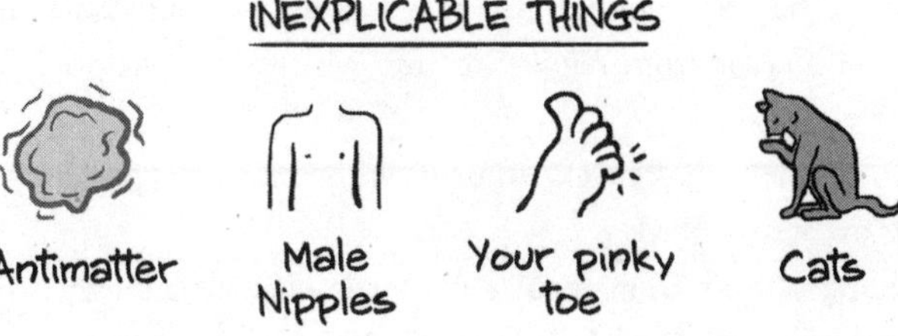

Neutral Matters

Does every particle have an antiparticle? So far, every particle that has electric charge has a distinct antiparticle. But the answer is not so clear for neutral particles.

For example, there is no distinct antiversion of the photon (which has no charge), i.e., an antiphoton. Some would say that the photon is its own antiparticle, which seems more like avoiding the question than answering it (i.e., does being your own best friend mean you have no friends?). The same is true of the *Z* boson and the gluon. You might notice that these are all particles that carry force, but the charged W particles also carry force

I'M ANTI-ME.
P
PHOTONS ARE THEIR OWN WORST ENEMY.

and they do have antiparticles. Why do some particles have antiparticles and other don't? We have no idea.

Physicists believe the neutrino (which has zero electric charge) probably has an antiparticle with opposite values of the charges associated with the weak nuclear force (called "hypercharge"). But neutrinos are mysterious little particles that are difficult to study, so it's possible that the neutrino is also its own antiparticle.

How Can We Study Antimatter?

It's fascinating to think that we could build antiobjects from antiparticles. That would be very cool but also educational: we could learn how antimatter is different from regular matter, which could help explain why antimatter exists.

Unfortunately, making experiments with antiobjects (made of antiparticles) is extremely hard.

Building objects from regular matter is difficult enough (to make a chocolate cake you need 10^{25} protons, 10^{25} electrons, and lots of love) without having to worry about your baking project exploding when it comes into contact with a single particle of normal matter.

In the case of antimatter, scientists have only recently succeeded in getting antiprotons and antielectrons to play together nicely enough in a lab to form antihydrogen. In 2010 scientists succeeded in creating a few hundred atoms of it and trapping them for about twenty minutes.[99] This is technically very impressive, but it is still not enough to answer all the questions we have about antimatter. Imagine how little you could learn about our universe if you were able to look at only a small number of hydrogen atoms for a few minutes.

So we are making really great progress, but we probably won't learn

99 In units of academic time, this is 1.0 coffee breaks.

more unless we get much better at making antimatter and storing it safely. Currently, we can produce only a few picograms of antimatter annually at CERN, which means it would take *millions of years* to make the equivalent of half a raisin of antimatter. And even then we would need to invent some form of no-contact container, perhaps by using electromagnetic fields.

IT'S ALL MADE OUT OF CHEESE!

Curious Matters

So we know a *few* things for certain about antimatter. We know that it exists, that it has the opposite charge of matter, and that when it comes together with matter it can annihilate and turn into light. We're not completely clueless.

But that understanding is dwarfed by the things that we *don't* know about antimatter. First, we don't know why antimatter exists. Is it a clue about the way matter is organized? Could there be other forms of matter? And while there seems to be a lot of symmetry between matter and antimatter, the universe definitely has some preference for matter.

All of these questions might make you wary of antimatter. It's obvious you don't want to touch it, but think about all the cool things we can learn from it.

For example, a huge question we still have is: Do antiparticles feel gravity the same way as matter particles do?

Even though we know antimatter exists and the current theory predicts that it feels gravity just like normal matter, we actually haven't been able to observe significant enough quantities of it to answer this basic

WHAT'S THE MATTER WITH ANTIMATTER?

question. Gravity is such a weak force that you need a very large number of particles to measure it. Antimatter is so rare and unstable that gravitational experiments are nearly impossible.

But what if antimatter feels gravity differently than regular matter? Remember that the defining feature of antiparticles is that their electromagnetic, weak, and strong force charges are reversed. Is it possible that antimatter particles also have their "gravity charge" reversed? Could it be that antimatter feels gravity in the *opposite* way? Imagine what would happen if this was true and that we somehow figure out how to create and harness antimaterials with this "antigravity" property. Those flying cars and antigravity boots you fantasized about as a kid might actually become a reality!

If *that* happens, we might want to change the name of this stuff from "antimatter" to "awesome matter."

WHO NEEDS MATH WHEN YOU HAVE ANTIGRAVITY BOOTS?

13.

What Happened to Chapter 13?

We have no idea.

14.

What Happened During the Big Bang?

And What Came before It?

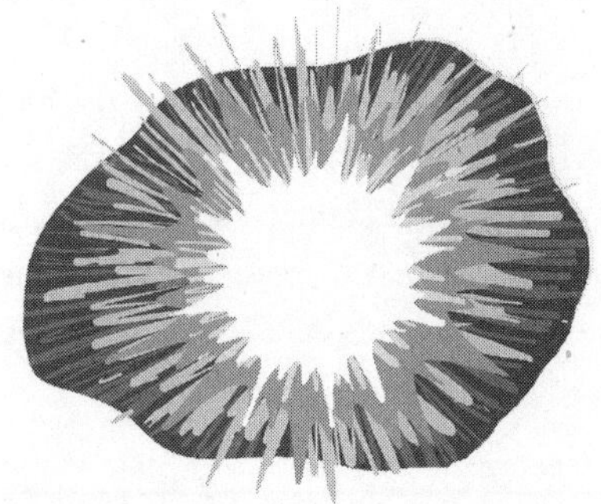

If someone told you that you were born under mysterious circumstances, wouldn't that pique your interest? If you were told that you suddenly appeared on Earth as a baby and nobody knew if you were grown in a test tube, assembled in a factory, or popped into existence by aliens, wouldn't you find that alarming?

Knowing where you came from and how you came to be is an integral part of your identity. The knowledge that you were conceived and born probably sits comfortably in the back of your mind, reassuring you that it is normal for you to be here and that you are part of some larger history.

But that is not the case with the universe.

Our universe came into existence about 14 billion years ago (we'll talk later about how we know this), and to say that it happened under mysterious circumstances is probably the mother of all understatements.

Scientists think they know what happened just *after* the universe was born—a huge expansive explosion called the Big Bang—but they know very little about the actual moment of birth, what caused it, and what (if anything) came before.

In this chapter, we'll talk about all the things we know, and don't know, about this extraordinary event. Spoiler alert: it probably wasn't grown in a test tube.

How Can We Know Anything About the Big Bang?

It's helpful in situations like these to remember the limits of science. Science is a pretty useful tool for answering many different kinds of questions, but it has its limitations. Namely, scientific theories have to make *testable predictions* that can be validated in experiments. For example, if you have a theory about your cat's behavior, you can test it by shooting at it with a Nerf gun and seeing how it reacts.

SCIENCE: IT'S GOOD FOR SOMETHING.

If a theory can't be tested with experiments, then it falls in the realm of philosophy or religion or pure speculation. For example, someone could suggest the theory that deep in space, between our galaxy and the Andromeda Galaxy, there floats a tiny pink stuffed kitten. This is a solid, physical theory, but our current technology makes it untestable. So for now, it's not a scientific idea, and believers in the Deep Space Kitten have to rely on faith or other arguments.

Theories have crossed the boundary from unscientific to scientific many times in history. The idea that matter is made of tiny atoms existed

THE DEEP SPACE KITTY
WATCHES OVER US ALL.

long before we had the technology to detect those atoms. Questions like this were converted from philosophy to science by creating new tools with greater power and insight.

Such is the case with the Big Bang.

Up until recently, it would have been the subject of pure speculation to talk about the early moments of the universe. After all, how do you study something that happened 14 billion years ago? And more important, how do you do experiments to verify your theories? It's not like we can rerun the Big Bang for our scientific convenience.

Fortunately for us, the Big Bang left a big mess. There are all kinds of clues and bits of rubble for us to analyze in detail. And in the last half century, our technology, mathematics, and physics theories have progressed to the point where we have started to move the question of what happened during the Big Bang to the scientific category. We can test theories about the Big Bang as long as they make predictions about things we can find in the rubble; that counts as a prediction even if the events happened a long time ago.

But just because we have this ability doesn't mean that we know everything about the Big Bang, especially about what came before it. To understand what we don't know about the Big Bang, let's first talk about what we *do* know.

What Do We Know about the Big Bang?

The idea for the Big Bang came in the early parts of the twentieth century when scientists discovered that all the galaxies we could see were moving away from us, which meant that the universe was expanding.

Cosmologists tried to make sense of this observation by playing with Einstein's new equations for general relativity, which describe how space and time and gravity work, and found that these equations could easily describe an expanding universe. But they also found something odd. If you project that expansion backward in time as far back as possible, then the equations predict something that is almost totally alien to our intuition: the entire universe contained in a single point, a *singularity*, where the mass is enormous, the volume zero, the density infinite, and the parking *impossible.*

That growth from a tiny small seed to the vast and grandiose universe we see today is what we call the Big Bang, the origin of our universe.

Most people who've heard of the Big Bang probably think of it as an explosion, similar to what happens when a bomb detonates. They imagine that before the Big Bang, all the matter in the universe was crammed into a very small volume, and that afterwards, all that matter flew outwards through space, leading to the universe we see today.

But if you find it hard to believe that everything that now exists was crammed into a single infinitesimal point and then exploded outward, you have a good point. The story of what happened during the Big Bang is *a lot* more complex than that, and it's full of mysteries to which we currently have no answers. Read on to find out what they are.

Big Mystery #1: Quantum Gravity

Let's start at the beginning. Does it make sense that our universe was once a single infinitesimal point? That all the stuff that exists today was once at the exact same location, squished down to zero volume? Actually, according to general relativity, it does make sense.

But general relativity was conceived and developed before it became clear that at the smallest distances our universe is a strange place populated by quantum objects that obey weird counterintuitive and probabilistic rules. The predictions of general relativity are expected to fail when masses get so dense that quantum mechanical effects become important. Like during the early moments of the universe when things really were squished down into incredibly small spaces.

Sometimes you can't take a theory all the way to its logical conclusion. Imagine if you measured how quickly your cats were growing over time and then tried to extrapolate their growth backward in time. If you only went by size, you might end up with the prediction that your pets were once infinitesimal kitten singularities or, if you ignore the physical boundaries completely, that they once had negative size. That would be . . . *cat*astrophic.

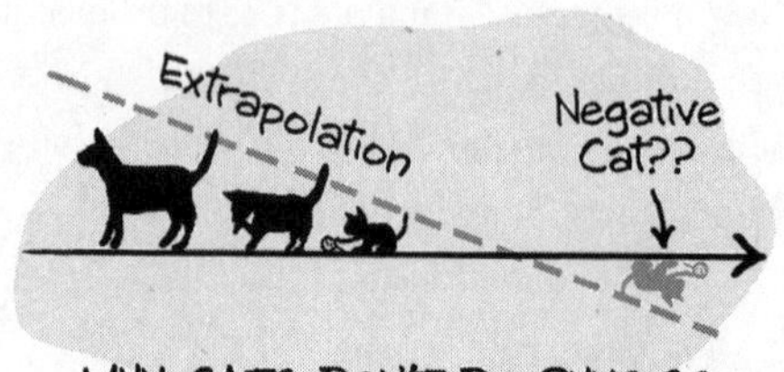

The same is true of general relativity and the Big Bang. Since we don't have a quantum theory of relativity, we don't really know how to calculate or predict what was happening in the very early universe. This means that the picture of the Big Bang starting with a singularity is probably not accurate; in those early moments quantum gravity effects dominated, but we have no idea how to describe them.

Big Mystery #2: The Universe Is Too Big

There is another problem with the simple view of the Big Bang as an explosion from a small original nugget. Even if the universe grew from a small infinitesimal point or a small blob of quantum blobbiness, there is still something that doesn't quite match what we see: the universe is bigger than it should be.

To understand this, let's first think about how much of the universe we can see. Beyond the book in your hands, the cat on your lap, the world outside the window, think about the distant stars. How far away could you see if you had a powerful telescope capable of catching the light straining to reach us from those distant stars? The answer depends on how *old* the universe is.

Seeing something means that you are catching photons that started their journey at the thing you are trying to see and then made their way to your eye (or your telescope). But because there's a limit to how fast photons can travel (they can travel only at the speed of light), then seeing something really far away means a lot of time has passed from the instant the photon was emitted to the moment you caught it.

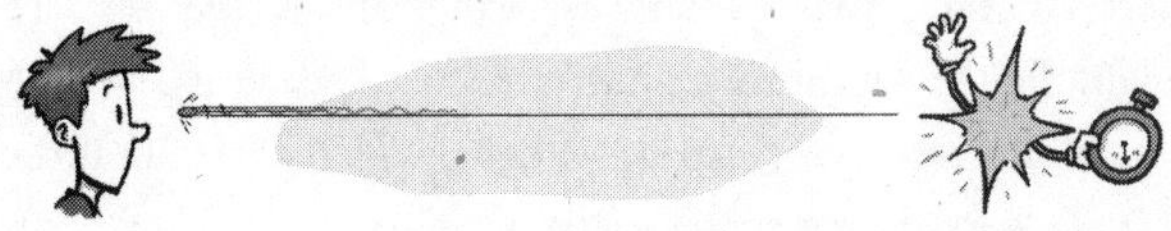

So how *far* you can see depends on how much *time* has passed since the universe started.

If the universe started five minutes ago, then the farthest you could see would be five minutes times the speed of light, or about 90 million kilometers.[100] That may sound like a lot, but it means you would be able to see only about as far as Mercury.

This is the "observable universe." Everything you can see has to be inside a sphere centered at your head whose radius is the distance that light can have traveled since the universe was born. If a point on the surface of that sphere sent you a photon at the earliest possible moment, it will only be arriving now; that is what defines the edge of our vision.

Light from stars, planets, and kittens outside that sphere will not yet have reached us, so no telescope can see them. Even a superbright supernova or a giant planet-size pink kitten would be invisible to us if it was outside that sphere. Oddly enough, this concept has returned us to our

100 Assuming there is no expansion of space itself—we'll get to that in a minute.

ancient place as the center of the observable universe except that we are each at the center of our own observable universes!

As more time passes, this sphere grows outward and we can see more of the universe. Each year we can see farther and farther out because we are allowing light from more distant objects to reach us. And this information is coming to us at the speed of light, which means that the edge of our vision is also growing at the speed of light.

But at the same time, everything in the universe is moving away from us, so there is a race going on between the edge of our vision and the targets of our telescope. How close is this race? The edge of our vision is growing at the speed of light, but the stuff in the universe can't travel through space faster than that (according to relativity).

So if all the things in the universe started from a tiny but finite quantum dot and are simply moving through space away from the Big Bang, our horizon should expand faster than the stars and kittens of the universe can move away from us, giving us a longer and longer view. Very quickly, if not already, our horizon would be larger than the entire universe.

What would that look like? When our horizon is bigger than the universe, it means that we can see beyond the point where there are no more stars (or were no stars, since what we see happened a long time ago). We'd be looking at a spot that had nothing: *an end to the stars*.

But in every direction we look, we see no end of stars. The universe is *still* bigger than our horizon, even though 14 billion years have passed since the beginning of the universe. Clearly, there is something not quite

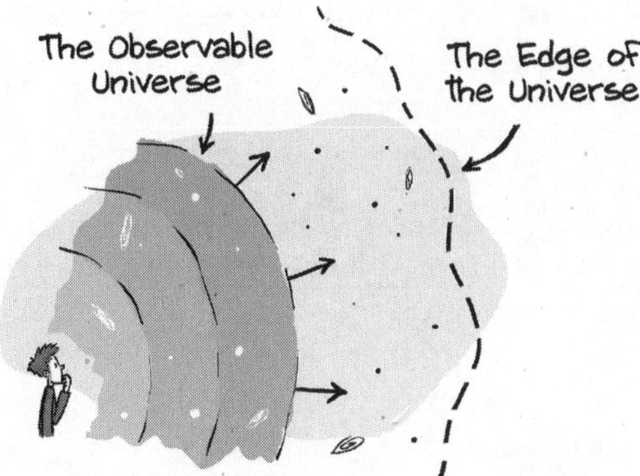

correct about this idea that everything in the universe started from a small blob and simply moved outward through static space.[101]

And it gets even worse.

Big Mystery #3: The Universe Is Too Smooth

There are other problems with the idea of everything in the universe simply moving away from a small starting point during the Big Bang. Namely, that the universe is too *smooth*.

As awesome and chaotic as the universe may seem to you, there is actually a kind of pervasive homogeneity or uniformity about it. And we can see this uniformity in the cosmic microwave background, the CMB from chapter 3.

To understand this, let's look at an example. Imagine that you are hungry (reading books about physics burns a lot of calories—tell your friends) and decide to heat up a pastry in your microwave oven. After a

101 This assumes the universe is finite. If you don't, then you avoid this problem because an infinite universe will always be bigger than what we can see, but then you'd have the problem of explaining how an infinite universe is created.

few minutes, as everyone knows, the center of your pastry will be really hot while the outer edges will be less hot.

Now imagine that you were inside the pastry taking the temperature of your surrounding microwaved goodness.

If you were standing at the center of the pastry, you would find that the temperature from all sides of you is the same.

But now imagine that you are standing just to the side of the center of the pastry. If you measured the temperature on the side closest to the center of the pastry, you would find it to be really hot. But if you measured the temperature in the other direction, toward the edge of the pastry, you would find the temperature to be cooler.

You can do the same thing with our universe from our little spot called Earth. We can measure the temperature of the CMB photons that are hitting the Earth on one side and compare that to the temperature of the photons hitting the Earth from the other side. And what we find is a little surprising: the temperature is the *same* (about 2.73 K) no matter which direction you look!

It seems unlikely that we are standing at the *exact* center of a microwave-reheated universe, so we can only conclude from our measurement that the entire universe is at the same even temperature. That is, the universe is more like a warm bath that's been sitting there invitingly for a while rather than a freshly nuked pastry.

A WARM BATH vs. THE UNIVERSE

	WARM BATH	THE UNIVERSE
HAS WATER IN IT	✔	✔
EVEN TEMPERATURE	✔	✔
CONTAINS RUBBER DUCKIES	✔	✔

To understand how this spells trouble for our simple Big Bang theory, we first have to understand what the photons from this cosmic microwave background really represent: they give us the earliest picture of the universe as a baby.

In its early days, the universe was much hotter and denser than it is today. Back then, the universe was too hot even for atoms to form, leaving all matter in a state of floating ions called plasma. Electrons whizzed around freely, having too much energy and too much fun to be committed to a single positive nucleus.

But as the universe cooled, there was a brief period when this ceased to be true: the temperature dropped enough that the charged plasma turned into neutral gas, and electrons started to orbit around protons to form atoms and elements. During this transition, the universe went from being *opaque* to being *transparent*.

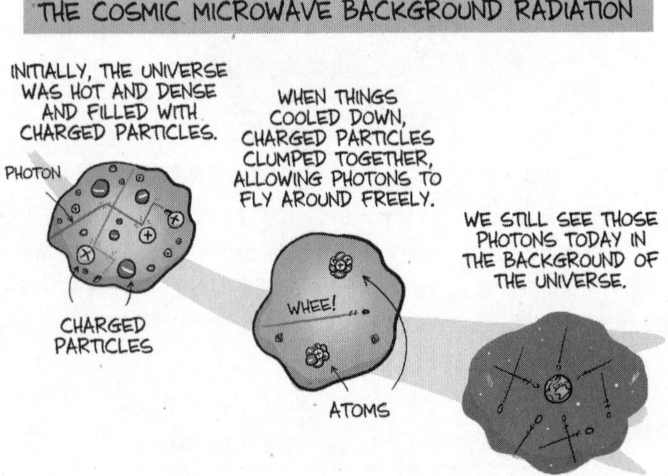

THE COSMIC MICROWAVE BACKGROUND RADIATION
INITIALLY, THE UNIVERSE WAS HOT AND DENSE AND FILLED WITH CHARGED PARTICLES.
PHOTON
CHARGED PARTICLES
WHEN THINGS COOLED DOWN, CHARGED PARTICLES CLUMPED TOGETHER, ALLOWING PHOTONS TO FLY AROUND FREELY.
WHEE!
ATOMS
WE STILL SEE THOSE PHOTONS TODAY IN THE BACKGROUND OF THE UNIVERSE.

Back in the plasma phase, photons couldn't get very far without bumping into the freely moving electrons and ions. But once the electrons and protons (and neutrons) formed neutral atoms, it was much rarer for photons to interact with them, so the photons could move around more freely. To the photons, the foggy universe suddenly became crystal clear. And because the universe has mostly gotten cooler since then, most of those photons *are still flying along untouched*.

These are the photons that we detect when we measure the cosmic microwave background radiation, and the curious thing is that the temperature of these photons seems to be the same everywhere.

No matter which direction you look, you see photons at the same energy. The CMB is very, very, very smooth. This is what you expect if something has had a long time to mix and equalize and balance out any hot spots. For example, this is what would happen if you left your pastry in the microwave to cool for a long time. Eventually, all the molecules would be roughly the same temperature.

But remember that the CMB photons are very old; they date back to just after the Big Bang, making them about 14 billion years old.[102] If you look in one direction in the sky, you are seeing photons that were created 14 billion years ago, very, very far away. If you look in the opposite direction, you see photons created the same distance away in the other direction.

How could these photons have the same energy if they are coming from opposite ends of the universe? How could they have had a chance to mix with one another and exchange energy in order to equalize? It seems these photons would have had to communicate faster than the speed of light in order to mix with one another and have the same temperature.

An Inflated Answer

So the universe is too big and too smooth for it to have come from a Big Bang in which everything simply moved through space starting from a small blob. If we had written this book thirty years ago, this might be one

102 They don't like to talk about it. Don't ask.

of the great mysteries. Today, there does exist a compelling but totally crazy-sounding explanation. Are you ready?

What if, a few moments after the universe was created, there was a period of about 0.000000000000000000000000000000001 seconds in which the fabric of *space-time itself* expanded by a factor of about 10,000,000,000,000,000,000,000,000—at a rate faster than the speed of light?[103]

Bam. Problems solved.

What? Does a *nearly instantaneous twenty-five order of magnitude* faster-than-light expansion of the fabric of space-time sound totally ridiculous and made up? If so, you're probably not a crazy physicist.

In fact, this is the solution physicists have come up with to explain why the universe is bigger than it should be and why it is at an even temperature. They call it (*drumroll*) "inflation." Okay, not the most awe inspiring of names. But the crazy thing is that it is probably true.

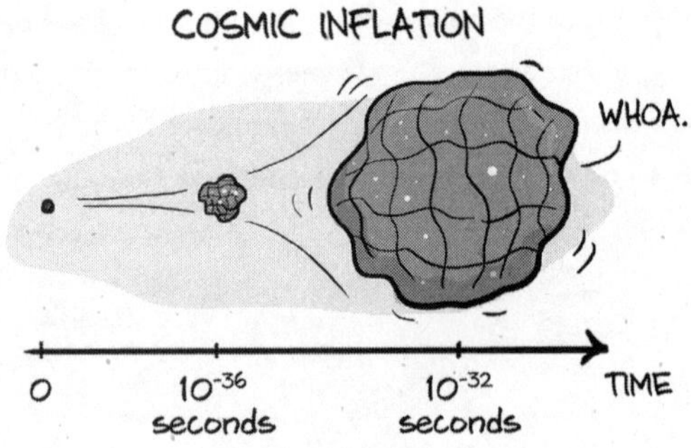

First, let's talk about how this solves the mystery of the universe being too big.

Remember that the problem was that the observable universe, which is growing at the speed of light, was still somehow smaller than the actual universe, which grows at a rate that should be slower than the speed of light. Well, inflation says that, just for a little bit, the universe expanded *faster* than the speed of light.

103 Faster than light here means the growth of new space, adding distance faster than light could cross it, not literal faster-than-light motion through space, which is impossible.

The things inside the universe kept obeying the cosmic speed limit (they didn't move *through* space faster than light), but according to inflation *space itself* did expand, making new space faster than light could traverse it.[104]

This is how a universe that starts from a tiny finite dot can now be so much bigger than the observable universe. During inflation, the universe blew past the horizon of the observable universe, pushing some things so far out that we haven't yet received the light they emit.

This expansion of space was very dramatic: the universe got bigger by a factor of more than 10^{25} in less than 10^{-32} seconds. After inflation ended, the universe kept expanding, first at a much slower rate and then more recently at a faster rate due to dark energy. Now the observable universe has a bit of a chance to catch up because it's still expanding at the speed of light. But how much of the universe is still way beyond the observable universe for us to see? We have no idea, but that's the topic for the next chapter.

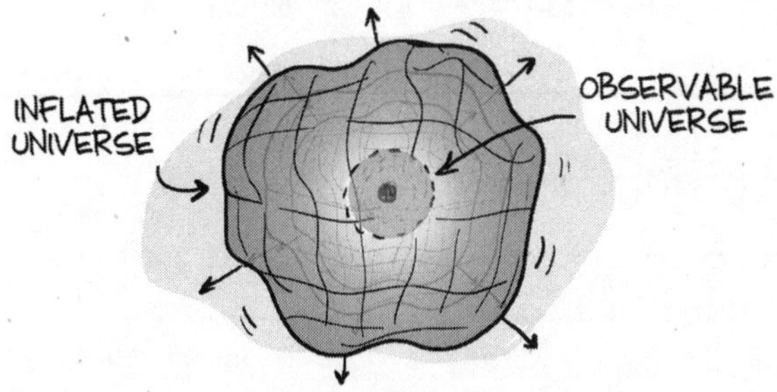

And how does inflation solve the problem of the too-smooth universe?

Solving the smooth-photon problem means finding a way for those early photons (the ones coming from different ends of the universe) to have mixed so they could even out in temperature; this can happen only if—at some time in the distant past—those photons were much closer to one another than the current rate of expansion predicts.

Inflation solves this problem by saying that the photons were indeed

104 Remember, space is a thing now, not just a backdrop. See chapter 7.

closer together at some point *before* the rapid expansion of space-time. Before inflation, the universe was small enough that there *was* time for all those photons to get to know each other and equilibrate, thus getting to the same temperature.

Once inflation happened, those photons got pulled apart to distances that to *us* seem impossibly far for them to have the same temperature. It just *seems* to us in the present day that they are too far apart to have talked to one another, but before inflation, they were plenty close together.

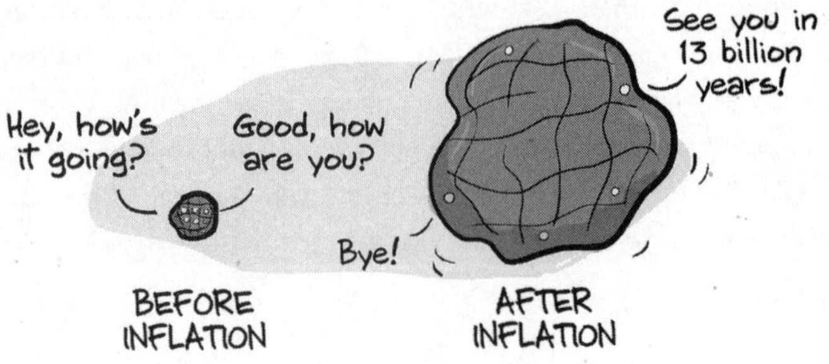

Are We Done?

This ridiculous and nearly instantaneous cosmic stretching called inflation makes everything make sense.

And the amazing thing is that it is *still* happening today. Not at the same absurd rate, but dark energy is still—right now—making new space.

Recently, this theory of inflation graduated from a crazy theory that makes all the math work to an experimentally supported (but not yet conclusively established) observation.[105]

How, you might ask, can we verify something that happened 14 billion years ago? Well, the theory of inflation predicts specific signatures in the tiny ripplings of the cosmic microwave background that we should see today, and some of those signatures seem to be present in experimental

105 More direct proof would be the observation of gravitational waves from inflation, but recent claims to have seen these were later revealed to be in error.

measurements of the CMB. Of course, this doesn't mean that we know inflation is real because there are other theories that also predict such wiggles, but it lends weight to it.

In fact, this is also how we know that the universe started about 14 billion years ago. From those ripples, we can estimate the proportions of matter, dark matter, and dark energy in our universe, and we can combine them in a model with the rate of expansion of the universe. This model tells us the age of the universe.

And there is another reason to like this idea. When we talked in chapter 7 about how space is a dynamic thing that is bent by the amount of energy and matter in the universe, we told you that it appears to be a strange coincidence that there is just the right amount of matter and energy in the universe so that space is almost flat. Well, inflation makes this less strange—the expansion of space tends to make space look flatter, the way a larger planet seems to have a flatter surface than a smaller planet. In fact, inflation *predicted* that space was flat before it was actually measured.

Great! The Big Bang is explained. Sure, we had to invent a crazy, momentary, and absurd expansion of space-time to make it all work, but experiments suggest that it actually (probably) happened.

But here is the thing: *we don't know what caused inflation.*

What could possibly cause the space-time of a small universe to suddenly and absurdly expand twenty-five orders of magnitude? We don't know. The mystery of the inflationary Big Bang is still a very deep one, and we are just getting a grip on what the right questions are.

WHO INFLATED THIS MYSTERY?

WARNING: Philosophy Ahead

Here we must depart the firm foundations of scientific theories and leap into the fuzzier world of the philosophical and metaphysical.

For now, most of the ideas we have about these questions are just that: untestable, crazy (yet exciting) ideas. Maybe in the future clever scientists will think of ways to test them and discover some totally shocking and bizarre truth about the origin of inflation and the Big Bang.

What Caused Inflation?

Do we really have no idea what caused inflation?

It turns out that physicists do have some ideas for what might have caused inflation. And the good news is that according to one of these ideas we don't need to invent any new cosmically powerful forces of nature, only a totally *new kind of substance*. No big deal.

Here is the idea: What if the early universe was filled with an unstable *new* kind of substance that causes space-time to expand rapidly?

See? That was easy. Now we have to answer only two simple questions:

1. How does this new kind of substance cause space-time to expand?
2. If this new substance existed, where is it now?

In theory it's possible for a different type of matter to cause space-time to expand in the same way that regular matter bends and distorts space-time when we talk about general relativity and gravity.

How would this work? Well, gravity is almost always an attractive force, pulling masses together. But there are certain properties of mass and energy that could have the effect of expanding space-time so that things are pushed apart instead of pulled in. Think of this as the fine print of general relativity. This property is the pressure component of the energy-momentum tensor of matter. This sounds technical, but it means that under certain conditions (negative pressure) substances can cause space to expand.

MASS AND ENERGY BEND SPACE, PULLING THINGS TOGETHER...
...COULD NEGATIVE "PRESSURE" CAUSE INFLATION?

Of course, this makes you wonder where this inflationary substance went and why inflation stopped. The answer is that this inflationary stuff is unstable: it decays or breaks down into regular matter eventually.

So the theory goes something like this: maybe the early universe was filled with something that has negative pressure, and this negative pressure pushed space-time to expand very, very rapidly. Eventually, this hypothetical inflationary stuff transformed into more familiar matter, ending the crazy expansion and resulting in an enormous hot universe filled with dense normal matter.

This theory seems crazy, but it would explain what caused inflation. And remember that inflation seemed like a crazy theory before it explained lots of things we didn't understand about the early moments of the universe.

Of course, we have no idea what this weird negative pressure stuff is, but the concept of it is not that absurd (by physics standards). Cosmically powerful repulsive forces that cause the universe to explode to an absurd degree became less absurd in the last few decades with the discovery of dark energy. We know that something called dark energy is causing our universe to expand faster and faster (see chapter 3), but as with the negative pressure stuff that may have caused inflation, we don't know what it is. Are they related? Again, we have no idea.

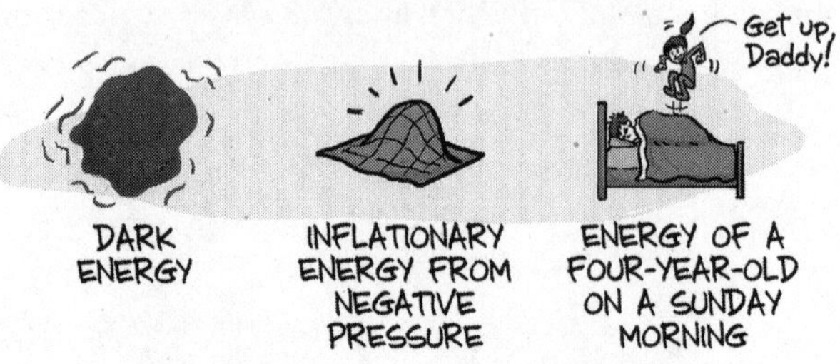

And What Happened before the Big Bang?

As mysterious as the circumstances around the Big Bang are, there's an even bigger mystery just on the other side of it. What caused the Big Bang? And what happened before it?

This question made sense when we thought about the Big Bang as a specific moment when the universe was a tiny dot, the clocks all read $t = 0$, and things started explosively from that first instant.

But now we have replaced the tiny dot with a fuzzy quantum blob (maybe small, maybe infinite), and the explosion has been replaced with inflation followed by dark-energy-fueled expansion. So the question still has meaning, but we have to first rephrase it in our new context. Instead of asking what came before the Big Bang, we should ask: Where did the quantum inflating blob come from?

Did that blob lead inevitably to a universe like ours, or could it have been different? Could the blob happen again? *Has it happened before?* The answer is that, as usual, we have no idea.

The exciting thing is that there is very likely an answer to these questions, and the evidence needed to reveal it might be within our grasp if only we had the tools. In the following pages we'll explore some possibilities about the origin of the universe that range from fairly simple ideas to theories that would seem outlandish even to dedicated science fiction readers.

1. Maybe the Answer Is That There Is No Answer?

Not every question has a satisfying answer because not every question is well posed. That might be the case for questions such as "What happens after you die?" because it depends on whether there is still a "you" after "you" die. Similarly, the question of "Why doesn't my cat love me?" might be ill posed because we don't even know if cats *can* love.

Even crisp mathematical questions can fall in this category. Stephen Hawking has suggested that asking "What came before the Big Bang?" is like asking "What is north of the North Pole?" At the North Pole, every direction you walk points south, and there is no more northness. This is just a feature of the geometry of the Earth. If space-time was created at the moment of the Big Bang, then it's possible that the geometry of space-time means that there is no satisfactory answer to the question of what came before (i.e., there is no "before").

EVERY CHILD KNOWS WHAT'S NORTH OF THE NORTH POLE.

As far as we can see, the universe seems to follow physical laws, and so even the creation of the Big Bang should be describable in such terms. But it's possible that from our vantage point inside space-time we don't have access to the information necessary to learn what came before it. Such a cataclysmic event might have destroyed any information about what happened before, leaving no evidence for us to discover. That is very unsatisfying, but there is no rule that all of science's answers will make us feel good.

2. Maybe It's Black Holes All the Way Down

A central question if we accept inflation is how the incredibly dense and compact inflationary stuff was created. When examining the universe for things that can create hyperdense pockets of matter, an obvious candidate is a black hole. Inside the event horizon of a black hole, matter is squeezed with intense gravitational pressure. Some physicists speculate that the strange negative pressure that caused inflation could have been formed inside a massive black hole.

In fact, you can take it a step further to suggest that our entire universe could exist within the event horizon of this mother of all black holes. Indeed, black holes in our universe might contain their own miniuniverses. These ideas are currently untestable. But they sound pretty awesome.

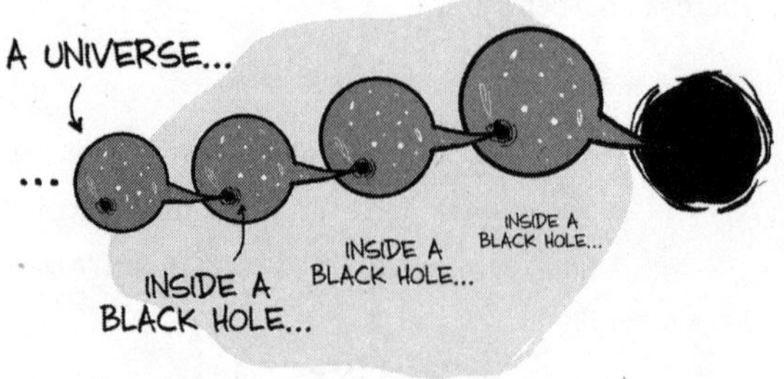
A UNIVERSE...
...
INSIDE A BLACK HOLE...
INSIDE A BLACK HOLE...
INSIDE A BLACK HOLE...

3. Maybe There's a Cycle

What if our Big Bang was just one of many? Maybe in the far future, dark energy and inflation will be reversed, causing a cosmic collapse called the Big Crunch. This crunch squeezes all the stars, planets, dark matter, and cats down into a tiny dense blob, which then triggers a *new* Big Bang. This cycle could potentially go on forever: *crunch, bang, crunch, bang, crunch* . . . There are some theoretical problems with this, however, involving the decrease in entropy of a crunching universe, but we are clueless enough about the arrow of time that there are potential solutions if you are willing to consider crazy ideas.

Of course, taking this idea from creative speculation to testable scientific hypothesis will be difficult. The conditions of the Big Bang will likely have destroyed any evidence of the previous iteration, which means we may never know the answer before the next Big Crunch comes around to crush us all to death.

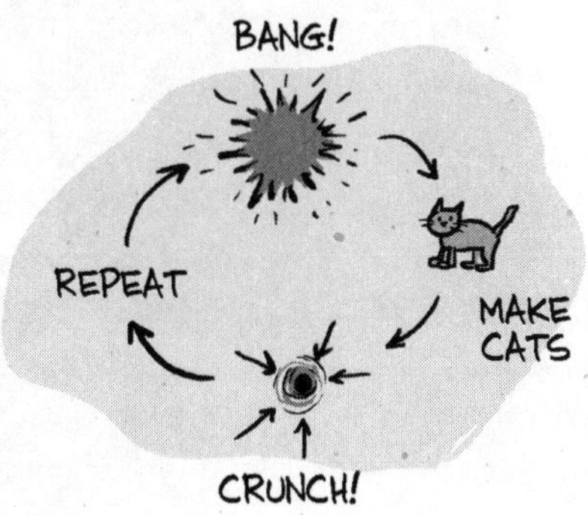

4. Maybe There Are Lots of Universes

Another possibility is that the weird stuff with negative pressure expands rapidly, and as it expands, it creates *more* of this weird stuff. And even though the weird stuff decays into normal matter, it's possible that it doesn't decay fast enough.

If more new weird stuff is created faster than it can decay into normal matter, then the result is that the universe will continue to inflate *forever*. Some parts of it will decay, but this will be overwhelmed by the creation of new inflationary stuff, which, if this theory is true, is continuing to inflate *right now*.

What happens in those spots where it decays? Each spot represents the *end* of the Big Bang in that part of space and the start of a slowly expanding universe of normal matter.

Each of these spots can form a "pocket universe" just like the one around us. Because the inflation continues forever, multiple universes are constantly being created. If inflation continues to create space faster than light can travel through it, then the inflationary stuff between the pocket universes will grow too quickly to allow these universes to ever interact with one another.

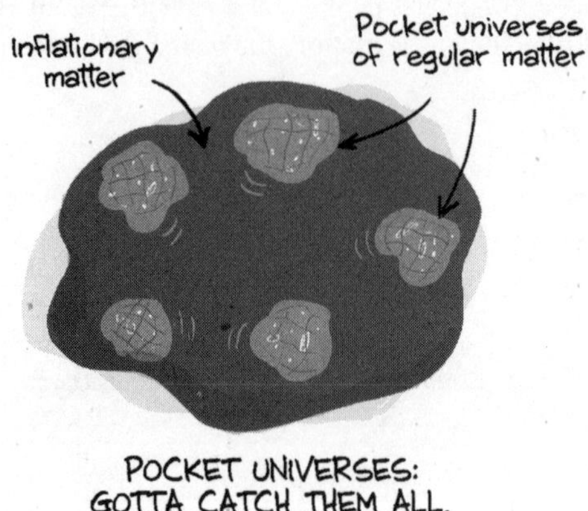

POCKET UNIVERSES: GOTTA CATCH THEM ALL.

What are these other pocket universes like? We certainly have no idea. It could be that each pocket universe is similar to ours, with the same laws of physics but slightly different random initial conditions, leading to structures similar to those we have here. If inflation has gone on forever and continues forever, that means an infinite number of pocket universes might exist.

Infinity is a very powerful concept because it means that every possible event will occur regardless of how unlikely. More than that, in an infinite number of universes an improbable event will happen *an infinite number of times* as long as the probability is not zero. If this theory is correct, that means that other universes can contain nearly identical copies of the Earth, including ones where a massive asteroid never wiped out the dinosaurs or one where the Viking colonization of North America was more successful and you are reading this book in Danish. Or one where your cat actually likes you.

The Big Finish

The fact that we have any clue whatsoever about the physics of the Big Bang is absolutely amazing. Imagine trying to reconstruct the circumstances of your birth if you didn't know anyone who was there at the time or if it had happened *14 billion years ago*.

On that time scale, our time here on the Earth is but the blink of an eye. But somehow, in that blink, we've managed to look at the universe around us and find evidence that takes us back to the beginning of time and the farthest reaches of the observable universe.

And as our time here grows beyond a blink, just imagine what else we will discover. Maybe we'll figure out what caused inflation and in the process learn about new types of matter or new properties of existing matter that we didn't know about before.

Or even more exciting, maybe one day our knowledge will punch through those early moments of the universe, and we'll be able to see what happened *before* the Big Bang. What will we find on the other side? Other universes floating in a vast ocean of inflationary stuff? Or another version of our universe heading toward a Big Crunch?

Today these questions are philosophical, but sometime in the future, they might become scientific, and our descendants, and their pet cats, will know the answers.

Today's philosophy questions are tomorrow's precision science experiments.

WHAT LIES BEYOND THE BIG BANG?
ANYTHING IS PUSS-IBLE.

15.

How BIG Is the Universe?

And Why Is It So Empty?

Climb to the peak of a remote mountain on a sunny day, and you'll be rewarded with a stunning vista. Unless there's already a Starbucks there, you'll get a solitary unobstructed view that stretches for miles and miles.

This feels impressive because—assuming you are not a billionaire with a penthouse apartment—the view you have out your window as you drink your morning coffee is probably measured in meters rather than miles. Maybe you're even close enough to your neighbor's building to be reading this book over her shoulder right now.

SO MUCH FOR WORD OF MOUTH.

But an even grander view is available every single night when you look up at the stars. This view lets you stare billions and billions of miles into space. Imagine each star as an island in the 3-D ocean of the universe. You can look across this immense sky and enjoy a dazzling spectacle of countless islands floating in space. Such a vision can give you vertigo if you remember you are perched on the tip of a tiny rocky island called Earth in this wide cosmic ocean.

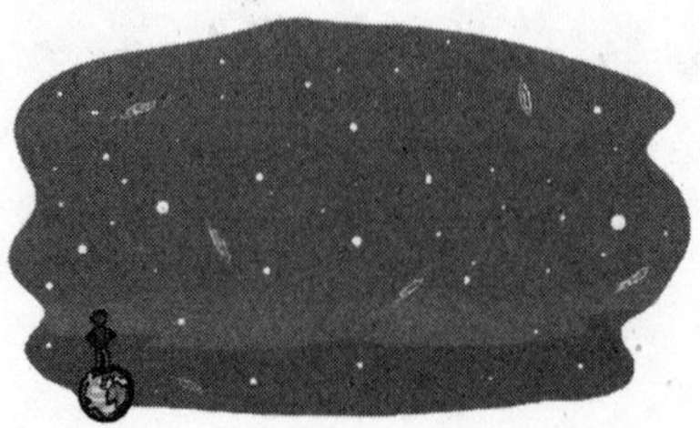

This view is possible because the universe is incredibly vast and mostly empty.

If stars were closer together, the night sky would be much brighter and going to sleep at night would be much harder. If the stars were much farther apart, the night sky would be depressingly dark and we would know a lot less about the rest of the universe.

Even worse, if space wasn't so transparent, this incredible view would be foggy and we would be deeply ignorant about our place in the universe. Happily, the kind of light that our Sun puts out and that our eyes are so good at seeing is pretty good at passing through the interstellar gas and dust. (Although infrared light and longer wavelengths are even better at this than visible light.)

So, fortunately, all of us (even the nontrillionaires) can see deep into space. But seeing is not understanding. Our ancestors stared at the same view and mostly got it totally wrong. In prehistoric times, even the richest among us had little clue as to the incredible knowledge that was washing over them. Today, thanks to telescopes and modern physics, we can look into space and understand our cosmic coordinates and the way that stars and galaxies are distributed.

But like our ancestors before us, we are probably still missing clues about the bigger picture, and our understanding only raises more questions: Are there more stars than we can see? How big is the universe? Can I still get a decent latte that far out?

In this chapter, we will tackle the biggest topic known to humankind: *the size and structure of the universe.*

You might want to hold on to something.

Our Address in the Cosmos

You are reading this book somewhere on Earth. Where exactly doesn't matter much in the grand scheme of things. Maybe you are sitting on your couch petting your hamster, swinging in a hammock in Aruba, or reading this on a toilet in a Starbucks somewhere. Even if you are a quadrillionaire floating above the Earth in your private space station, these details are irrelevant on the vast scale of the universe.

This third planet and its seven[106] sister planets follow the Sun as it orbits the center of our galaxy, which is a massive spiral disc with several arms swirling out from a bright central hub. We live about halfway down one arm of the Milky Way Galaxy. Our Sun is one of about 100 billion stars in our galaxy and is neither one of the oldest or youngest, nor the largest or smallest. Goldilocks would find it to be just right. When you look at the stars at night, you are mostly seeing the other stars in our arm of the galaxy, which are nearby on a cosmic scale. And on a clear night, if you are far from the light pollution of corporate coffee shops, you can see far enough to spot the disc of the rest of the galaxy. This appears as a wide swath of fuzzy stars so numerous and dense that they look like someone poured milk across the sky (hence the name). Almost everything you see in the night sky is part of our galaxy because those are the brightest (and closest) objects.

106 Screw Pluto.

The rest of the universe is mostly dotted with galaxies; there's no evidence that there are lone stars floating between galaxies. This is fairly new information; as recently as one hundred years ago, astronomers thought that stars were sprinkled evenly throughout space. They had no idea that stars clustered together into galaxies until they built powerful-enough telescopes to notice what those blurry distant objects actually were. What a revelation that must have been, to discover that our galaxy, which at the time seemed like an entire universe unto itself, was just one of *billions and billions* of galaxies we can see in the cosmos. It follows the discovery that our world is not the only planet in the universe and that our Sun is one of many, many stars. In each case, the scale of our unimportance grows by leaps and bounds.

Fairly recently we learned that the galaxies themselves are not distributed evenly throughout the universe. They tend to clump together into loose groups[107] and clusters, which themselves group together into

107 Ours is cleverly named the "Local Group."

massive superclusters, each with dozens of clusters. Our supercluster weighs in at about 10^{15} times the mass of our Sun. Heavy stuff.

So far, up to the scale of galactic superclusters, the structure of the universe is very hierarchical: moons orbit planets, planets orbit stars, stars orbit the center of galaxies, galaxies move around the center of their clusters, and the clusters zoom around the centers of superclusters. The strange thing is that it ends there. Superclusters don't form megaclusters, superduperclusters, or uberclusters, but instead they do something much more surprising: they form sheets and filaments hundreds of millions of light-years across and tens of millions of light-years thin. These sheets of superclusters are impossibly vast structures, and they curve around to form irregular bubbles and strands that surround huge empty cosmic voids in which there are no superclusters or galaxies and very few stars, moons, or quintillionaires.

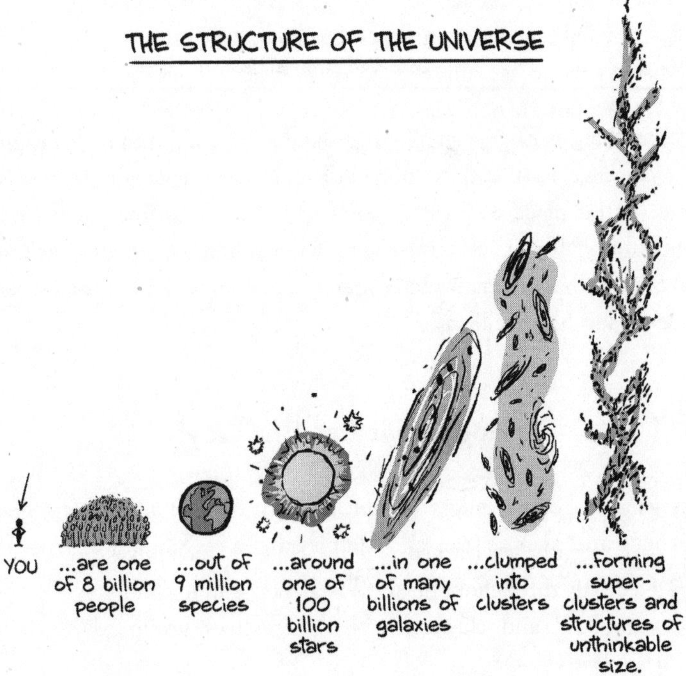
THE STRUCTURE OF THE UNIVERSE
YOU
...are one of 8 billion people
...out of 9 million species
...around one of 100 billion stars
...in one of many billions of galaxies
...clumped into clusters
...forming super-clusters and structures of unthinkable size.

This organization of superclusters is the largest known structure in the universe. If you continue to zoom out, you see the same basic pattern of stars-galaxies-clusters-superclusters-sheets repeating elsewhere, but no larger-scale structure is formed. The bubbles of supercluster sheets don't form into interesting complex megastructures. Like random Lego pieces on the floor, they are spread evenly across the cosmos. Why does the pattern end at this scale? Where do the supercluster bubbles come from? Why is the universe so uniform at this level?

DON'T STEP ON GALACTIC SUPERCLUSTERS

One thing is clear: compared to these scales, we are pretty insignificant. We have no special location in the universe; our cosmic address isn't some central place of great importance, like the cosmic equivalent of Manhattan.[108] And in a universe with many billions of galaxies, each with 100 billion stars, it remains to be seen if we are even that unusual when it comes to life and intelligence.

How Did It Get This Way?

Our galactic address might be old news to a well-educated and good-looking reader such as yourself.[109] But it raises a very interesting question: Why do we have this structure at all?

108 At best, we are in Poughkeepsie.

109 Did you lose some weight? You look great!

It's not hard to imagine a universe in which the stuff in it is arranged differently. For example, why are the stars not all gathered into a single megagalaxy? Or why is each galaxy not just a single star with a ridiculous number of planets around it? Or why have galaxies at all? Why couldn't we have a universe in which stars are evenly distributed, like dust particles floating inside an old room?

Or why have *any* structure at all? Imagine if in its first moment the universe was totally even and symmetric, with the same density of particles everywhere in every direction. What kind of universe would we get then? If the universe was infinite and smooth, then each individual particle would feel the same gravitational attraction in every direction, which means none of the particles would be compelled to move in any direction. The particles would never clump together and the universe would be frozen. And if the universe was finite but still smooth, then every

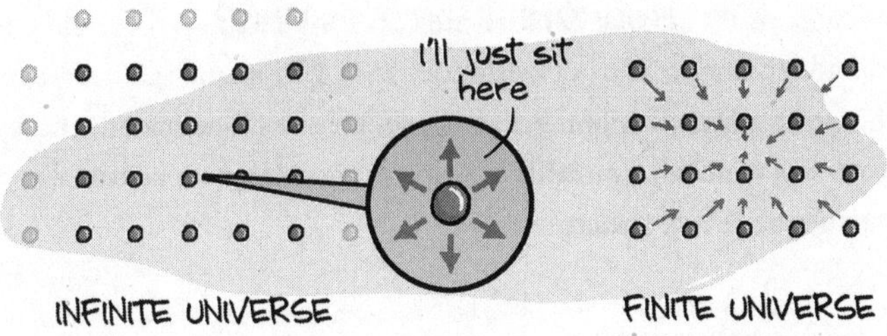

particle would be attracted to the same spot: the center of mass of the universe.[110]

In either case, you wouldn't get any local clumping or structure at all; this universe would be bland and smooth or clumped into one spot for its entire life, like a beige suburban café.

It turns out physicists have a pretty good story for how we ended up in a nonbland universe full of structure. Here's the theory: small quantum fluctuations in the early universe were stretched by the rapid expansion of space-time (i.e., inflation) into huge enormous wrinkles that seeded the formation of stars and galaxies by gravity, which was aided by dark matter; and at some point in there, dark energy started stretching space out even farther.

Phew. We said it was a good story, not an easy one.

You see, in order to have any structure in today's grown-up universe, you need *some* kind of clumpiness back in the universe's irresponsible youth.[111] As soon as you create the tiniest little clump with more mass than the rest, you form a local hot spot of gravity that can pull more and more atoms to it and away from the gravitational force of all the other atoms.

For example, imagine a city in which the Starbucks are spread equally distant from one another. Each coffee drinker will feel the delicious-smelling pull of those cafés closest to her, but since they are equally distant, she will be frozen in indecision forever. If, however, tiny fluctuations in the coffee-brewing process mean that one café had a stronger aroma, then it would attract more customers, leading to more Starbucks being opened across the street, which attracts more customers, leading to more Starbucks being opened, etc. This feedback loop creates a cascade, and pretty soon you have Starbucks stores opening inside of other Starbucks stores and leading to Starbucks singularities. But it can't get started without that initial hot spot. In the early pre-Starbucks universe, the first deviations from smoothness are absolutely crucial for creating today's arrangement of stars and galaxies.

110 It's also possible that the universe is finite but has no center if space is curved. Think of the finite surface of a sphere, which has no center.

111 The universe was out of control early on. Literally.

So what caused these first deviations from smoothness in our baby universe? The only mechanism that we know of that can accomplish this is the randomness of quantum mechanics.

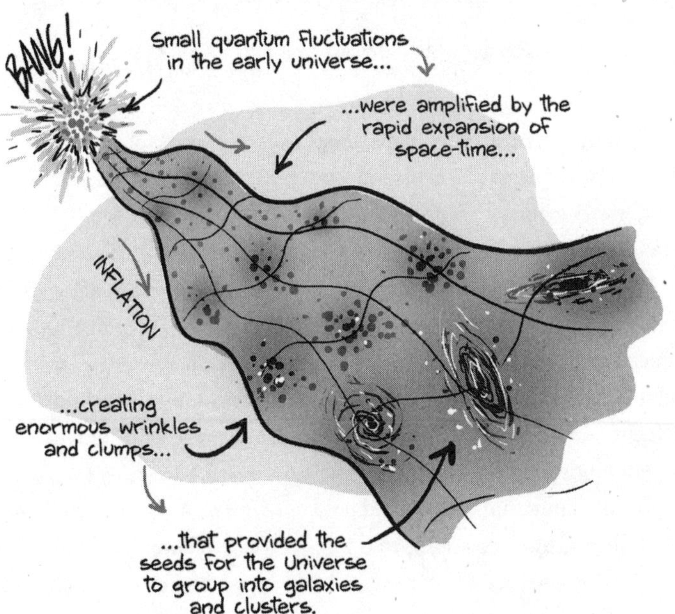
HOW QUANTUM FLUCTUATIONS GAVE THE UNIVERSE STRUCTURE:
BANG!
Small quantum fluctuations in the early universe...
...were amplified by the rapid expansion of space-time...
INFLATION
...creating enormous wrinkles and clumps...
...that provided the seeds for the Universe to group into galaxies and clusters.

This is not speculation—this is something that has been observed. Recall that we have a baby picture of the universe from the cosmic microwave background, which shows us what the universe looked like the moment it cooled from a hot, charged plasma to mostly neutral gas. In that image, we see that the universe was smooth but not perfectly smooth. It has tiny ripples that represent the quantum fluctuations of the early universe.

During the Big Bang, inflation stretched space tremendously and blew up those tiny ripples into huge wrinkles in the fabric of space and time.

These wrinkles in space-time then created the clumps and hot spots of gravity that later led to more complex structures.

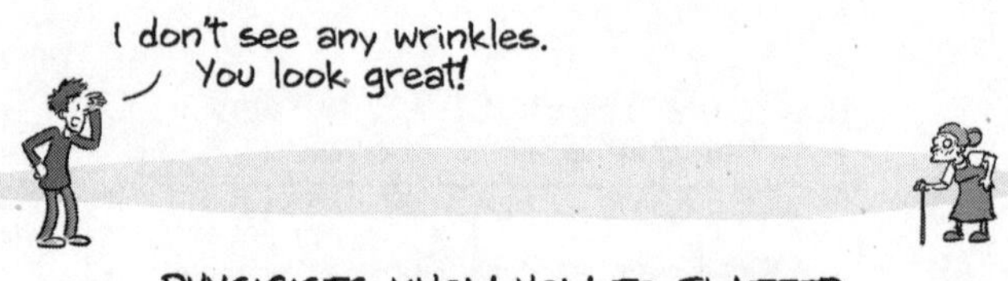

I don't see any wrinkles. You look great!
PHYSICISTS KNOW HOW TO FLATTER

To summarize, random rolls of nature's dice at the quantum level were blown up by the rapid expansion of space, which led directly to everything we see today. Without inflation, the universe would look a lot different.

Physicists suspect that the reason that there are no structures bigger than the sheets and bubbles of superclusters is that there just hasn't been enough time for gravity to pull things together and form more structure. In fact, there are parts of the universe today that are only just now starting to feel one another gravitationally because the effects of gravity are also limited by the speed of light.

What about the future? If dark energy was not expanding the universe, then gravity would keep doing its job of clumping things together and making ever-larger shapes and structure. But dark energy can't be denied. So we have two competing effects: having enough time for gravity to clump things into massive shapes but not so much time that dark

energy has pulled them apart. At the moment, these two effects seem to be perfectly balanced, which means we live in the perfect age to see the largest structures the universe will ever know.

Can that be right? Is it just coincidence that we live in the Ozymandias Age of the universe?[112] Anytime we believe that we live in a special place (e.g., the Earth as the center of the universe) or a special time (e.g., six thousand years after the creation of the universe), we should be extra careful that we're not just stroking our fragile egos.

It *seems* like we live in a special moment given our current understanding. But the truth is that we don't know for sure because we can't confidently predict the future of dark energy. If it continues pulling the universe apart, then there won't be time for galaxies and superclusters to pull together into more interesting structures. But if dark energy changes course, then gravity has a chance to pull things together and form new kinds of structures that we don't even have names for yet! Check back in 5 billion years for an update.

Gravity Versus Pressure

So the fact that we have *any* structure at all—rather than perfect smoothness—is due to the quantum fluctuations that created the first wrinkles, which were then blown out of proportion by inflation, creating the seeds that led to our current universe. But how do those seeds turn into the planets, stars, and galaxies that we see? The answer is a balancing act between two powerful effects: gravity and pressure.

112 "Look on my large-scale superclusters, ye mighty, and despair!"

Around 400,000 years after the Big Bang, the universe was a big blob of hot, neutral gas with a few little wrinkles in it. That's when gravity started to do its thing.

The fact that everything was *neutral* is very important. All of the other forces were approximately balanced at this point. The strong force grouped the quarks into protons and neutrons. Electromagnetism pulled protons and electrons together to make neutral atoms. But gravity can't be balanced or neutralized. It's also very patient: over millions and billions of years, it pulled those wrinkles together into denser and denser clumps.

AFTER ALL THE OTHER FORCES WERE BALANCED, GRAVITY GOT TO WORK.

But the universe has been around a long time, and you might wonder why gravity hasn't pulled everything back together into a big blob: either a massive star or a huge black hole or even a megagalaxy? It turns out that there is just enough matter and energy in the universe for gravity to make space "flat"—not bent enough to pull everything back together. And remember that dark energy is *expanding space itself*, so the net result is that things are getting further apart at large scales.

But even if gravity can't win this cosmic tug-of-war, it still scored little local victories. The wisps of gas and dust that were made from the original wrinkles got pulled together into bigger and bigger clumps even if those clumps were spread out through the universe.

What happens when gravity pulls a clump of gas and dust together? It depends on how big the clump is.

If you have a small blob of mass, then you only have enough gravity

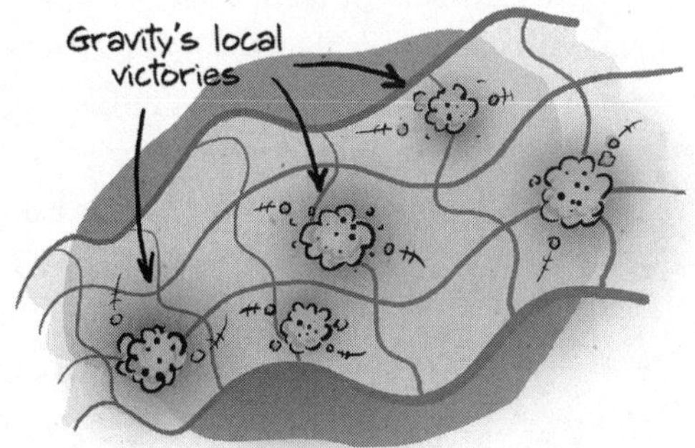

to form something like an asteroid or a big rock. Maybe a Frappuccino. The reason the rock or your venti beverage doesn't get collapsed into a tiny dot by gravity is that it has some internal pressure from the nongravitational forces. The atoms of a rock don't like to get squeezed together too tightly (Ever tried to squeeze a rock into a diamond? Not easy) and they resist. What you end up with is a balance between the squeezing of gravity and the internal pressure of the rock.

If you have a bigger mass, say enough to form a planet the size of Earth, the gravitational forces are strong enough to compress the rock and metals of the center into molten lava. The reason the center of the Earth is hot and liquid is due entirely to gravity. The next time you scoff at the weakness of gravity, ask yourself if you are capable of squishing a rock into hot lava.

That's what I thought.

If you get a big enough blob of matter, the gravitational forces can create a plasma hot enough to turn the blob into a star. Stars are essentially fusion bombs that are continually exploding; the only thing that contains them is their gravity. Gravity may be weak, but gather enough mass together and it can contain continuously exploding nuclear bombs for billions of years. The reason these stars don't immediately collapse into denser objects is also their pressure. Once they burn their fuel and can no longer provide the pressure to resist the relentless pull of gravity, some stars collapse into black holes.

This balance between gravity and pressure plays out for inert rocks, molten-lava-centered planets, and barely contained fusion-powered stars. It also explains why we have stars gathered into galaxies rather than just stars or black holes randomly sprinkled through the universe.

Remember that most of the mass in the universe is not the kind that forms planets and stars and coffee beans: about 80 percent of the mass (27 percent of the total energy) is in the form of dark matter. Dark matter might have some interactions we don't know about, but we are certain that its mass contributes to gravitational effects. Since it doesn't have electromagnetic or strong-force interactions, however, it doesn't have the same kind of pressure that resists gravity. So it clumps together in the same way that normal matter does, but it continues clumping, forming enormous halos. Wherever dark matter forms a halo, normal matter is pulled in by the large gravitational attraction. In fact, it's currently believed that dark matter is responsible for galaxies forming early in the history of the universe. In a universe

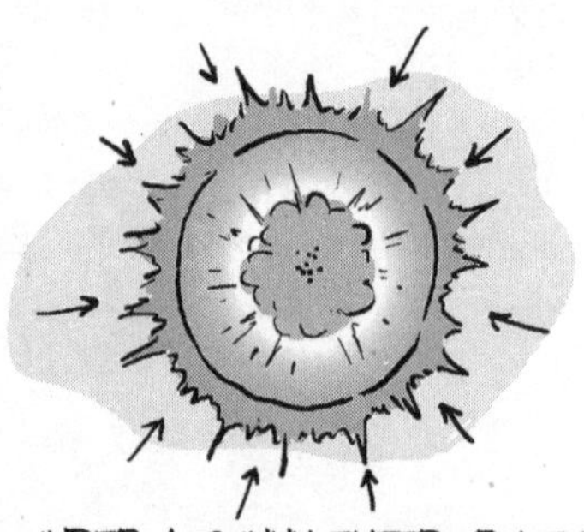
UNDER A SUNNY EXTERIOR LIES AN EXPLOSIVE PERSONALITY.

without dark matter, it would take many more billions of years for the first galaxies to form. Instead, we see galaxies forming only a few hundred million years after the Big Bang, thanks to the invisible hand of dark matter's gravity.

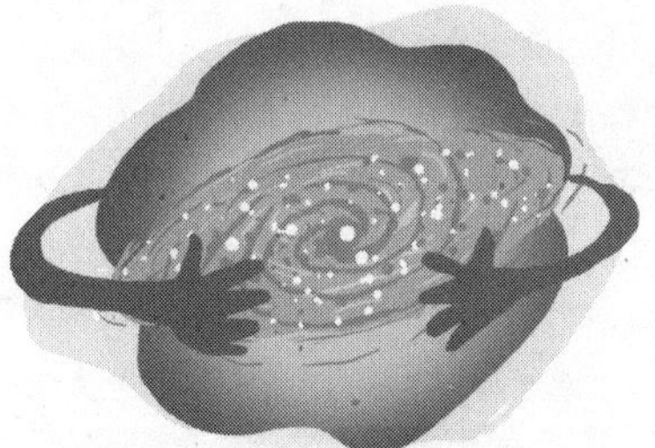

DARK MATTER JUST WANTS SOME BLING.

Galaxies are also pulled together by gravity, but they resist a total collapse into a massive black hole by various kinds of pressures, depending on the galaxy. Spiral galaxies don't collapse because they are spinning very fast, and the resulting angular momentum effectively keeps all the stars apart. This is also the reason dark matter doesn't collapse into denser clumps. The velocity and angular momentum of the dark matter particles make it difficult for gravity to pull it together.

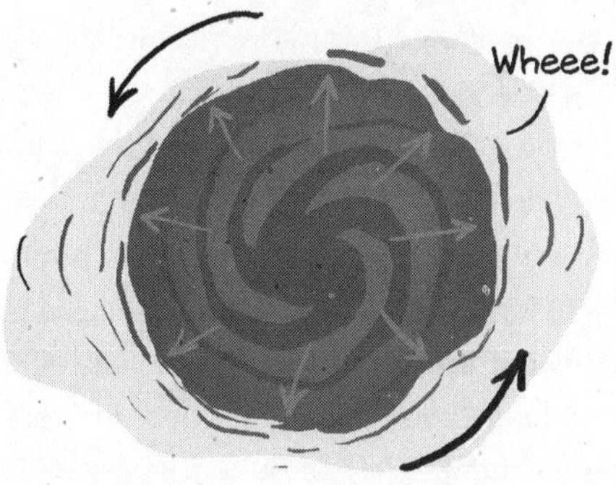

And so we end up in a universe filled with vast sheets and bubble structures made up of superclusters of clusters of galaxies, each with hundreds of billions of stars swirling around black holes and populated with dust and gas and planets. And on at least one of those planets are human beings looking out at the stars and pondering their existence.

But how far out does this go?

Do these sheets and bubbles of enormous size go on forever? Or is all the matter in the universe more like an island or a continent with edges that border on nothingness or infinity?

Just *how big* is the universe?

The Size of the Universe

If we could somehow drink an octo espresso and zoom around the universe infinitely fast, we would learn a lot about how things are organized, and more important, we would learn how far out things go.

Unfortunately, the largest espresso size offered at most coffee shops is a quad,[113] and the universe has a hard limit on how fast we can scoot around taking pictures. That means that until we develop warp drives we have to try to answer these questions using only the information that comes to the Earth from the Big Out There.

Light is screaming toward us carrying beautiful pictures of the strangeness of the universe, but it's only had 13.8 billion years to do it. That means that beyond that distance any object is *invisible* to us. There could be galaxy-size blue dragons cavorting and spitting just beyond our view,

113 They give us a funny look if we order two quad espressos.

and we would have no idea. Of course, nothing suggests that these dragons exist, but what are the chances that whatever is out there over the edge of our vision is exactly like the things we see around us? Nature is no stranger to bizarre and surprising revelations.

This sphere that stretches out to our horizon, called the observable universe, is very large. While we can't see what is outside that sphere, we can think precisely about how large it is. Consider some possibilities:

a. Since nothing can travel faster than the speed of light, the observable universe must be the age of the universe times the speed of light, or 13.8 billion light-years in every direction.

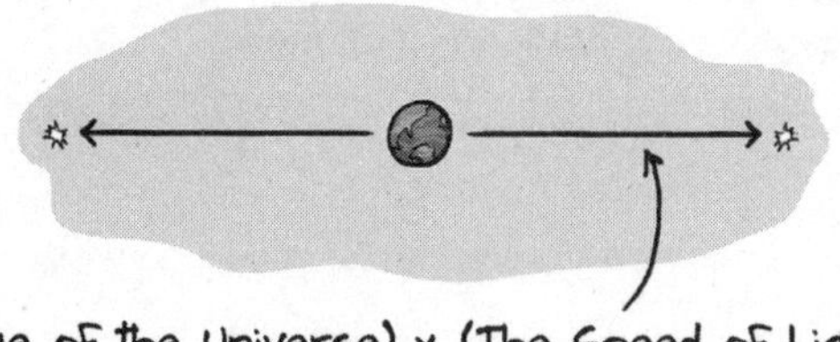

b. Since space itself is a thing that can expand faster than the speed of light (and has), we can see things that used to be inside our horizon but are now past it, up to about 46.5 billion light-years in every direction.

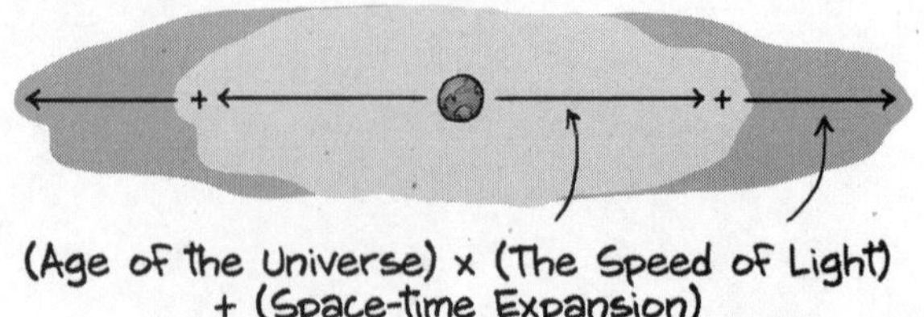

c. The observable universe is the distance between the two most far-flung Starbucks, currently unknown to science due to the rapid construction of new outposts.

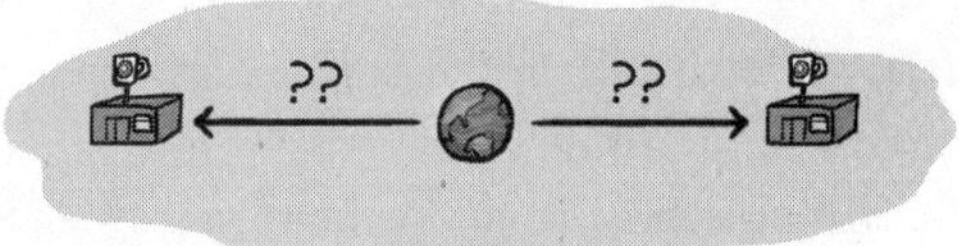

The correct answer is (b). Thanks to the expansion of space, we can see things that *used* to be closer to us than they are now. So the observable universe is much larger than the speed of light times the age of the universe. This comprises the universe that we can see today.

The good news is that we can see a lot, approximately 10^{80} to 10^{90} particles in sextillions of stars in many billions of galaxies. The other good news is that every year our observable universe—our visible

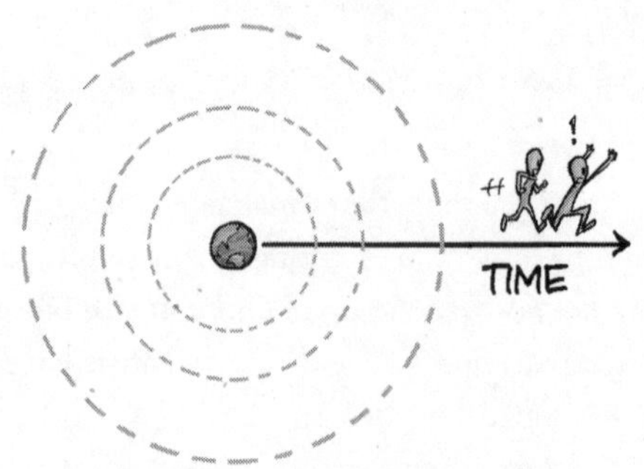

horizon—grows by at least one light-year without any effort on our part.[114] And thanks to the power of mathematics, the volume of the observable universe grows even quicker because the slice of space that is added every year has a larger volume than the previous year's slice, which means the number of galaxies with gorgeous mountains you will never visit is becoming a number we can hardly understand.

But it's not that simple. Things are moving through space away from us, and *at the same time,* space itself is expanding. There are objects whose distance from us is growing so fast that light from them will *never reach us*. In other words, the observable universe may never catch up to the actual universe, which means we might never see the full extent of everything that is.

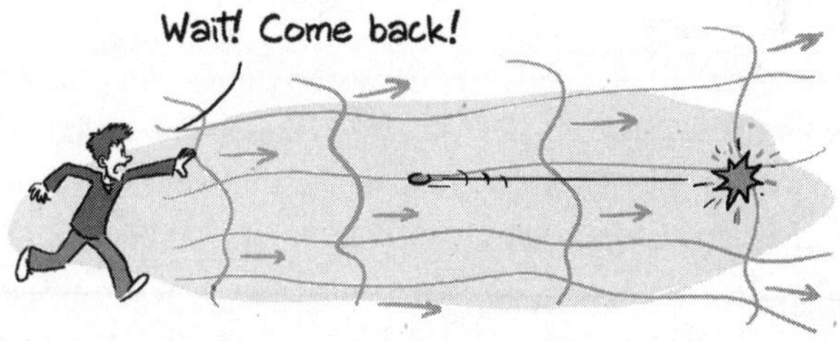

The bad news is that we don't know for sure how far out the universe goes. In fact, we may *never* know, which is worse news for those of you would-be cosmic cartographers.

Let's Just Guess

How big is the entire universe? There are a few possibilities.

A Finite Universe in Infinite Space

One possibility is that the universe is finite in size but grew past our horizon due to the expansion of space. Some scientists have run with this

114 Depending on the rate of expansion of space, which is currently greater than zero.

possibility and tried to estimate the size of the stuff in the universe by making some reasonable-sounding assumptions, such as:

- Before inflation, the size of the universe was approximately equal to the speed of light times its age, since space had not yet done any stretching.
- The number of particles in the universe is pretty large.
- Nobody can actually think about numbers bigger than 10^{20}, so you can pretty much guess whatever you want.

Take these assumptions and combine them with our present understanding of how much space stretched during the Big Bang and how much it is stretching now due to dark energy, and you can get an estimate of the size of the whole universe.

But depending on the nature of your assumptions, your answers will vary by more than a factor of 10^{20}. If that makes it sounds like the issue is far from settled, you are right. If someone told you that your house was somewhere between 2,000 and 100,000,000,000,000,000,000,000 square feet in size, you would correctly assume they were mostly guessing. Even if you can swallow the unjustified assumption that the amount of stuff in the universe is finite, we still have no idea how large the universe is.

THE UNIVERSE IS ABOUT THIS LONG...
...PLUS OR MINUS 10 SEXTILLION PERCENT.

Despite all this uncertainty, there are some scenarios under which we might be able to figure out the size of the universe.

A Finite Universe in Finite Space

If the shape of the universe is curved, it may be that space is like the surface of a sphere but in three (or more) dimensions. In that case, space itself is finite. It loops around on itself such that traveling in one direction would eventually lead you back to where you started. As startling as that would be, at least we'd know that the universe is finite, not infinite.

"TECHNICALLY, INFINITY IS FINITE," IS A REAL THING PHYSICISTS SAY.

But in this brain-twisting scenario, light traveling through such a universe would also loop around (assuming the loop is small enough) and might pass by the Earth more than once in its travels. This is something we could actually see! You would notice it by seeing the same objects in the sky multiple times, once for each time the light loops around.[115] Unfortunately, scientists have looked for such effects both in the structure of galaxies and in the CMB, but they have found no evidence for it. That means that *if* the universe is finite and looped, it must be bigger than what we can see.

Twinkle twinkle little star
how many times am
I seeing you tonight?

115 This is different from seeing duplicate objects in the sky due to gravitational lensing, which also distorts the objects. In this case, multiple versions would appear undistorted.

An Infinite Universe

It is also totally possible that space is infinite, and it is filled with an infinite amount of matter and energy. This is a mind-bending possibility because infinity is a strange concept. It means that anything that has any chance of happening (no matter how unlikely as long as the probability isn't zero) is happening somewhere in the universe. In an infinite universe, there is someone out there who looks like you and is reading a version of this book that is printed on polka-dotted sailcloth. There is a planet of blue dragons, all named Samuel, who keep getting one another mixed up. Think these scenarios sound unlikely? You're right. But in an infinite universe, anything that *can* happen *does* happen. To figure out how often something happens in an infinite universe, you multiply its probability by infinity. So as long as the possibility is nonzero, it's going to happen. And not only does it happen, it happens an infinite number of times. There would be an infinite number of planets with confused blue dragons. Mind: boggled.

Is the book any good?
In an infinite Universe, there is a version of this book that's really good!

But how can an infinite universe be consistent with what we see? Can a universe be infinite *and* expand from a Big Bang? Yes, but only if you don't assume the Big Bang started from a single spot. Imagine a Big Bang that happened *everywhere at the same time*. This is difficult to imagine

without splattering your brains all over the person next to you as you read this, but it's also totally consistent with what we observe. In such a universe, the Big Bang exploded *everywhere all at once*.

Which of these scenarios (finite matter in infinite space, finite matter in finite space, infinite matter in infinite space) is our reality? We have no idea.

And Why Is the Universe So Empty?

Another big mystery about the structure of the universe is: Why is the universe as *empty* as it is? Why are the stars and galaxies not closer together—or farther apart?

To give you some perspective, our solar system is about 9 billion kilometers wide, but the nearest star is about 40,000 billion kilometers away. And our galaxy is about 100,000 light-years wide, but the nearest galaxy, the Andromeda Galaxy, is about 2,500,000 light-years away.

However big space is and whatever shape it has, there seems to be plenty of room to have things closer together. It's not as if some cosmic parent had to separate all of the stars and galaxies because they were squabbling in the backseat.

Don't make me turn this Big Bang around! I'm serious!

Luckily, emptiness is a matter of perspective, and we can divide this question into two different questions:

Why can't we move faster than the speed of light?

and

Why did space expand during the Big Bang, and why is it still expanding today?

The speed of light is the cosmic yardstick that defines what we mean by "close" and "far." If the speed of light was much, much faster, then we would be able to see farther and travel more quickly, and things would not seem so far away. If the speed of light was much slower, our distant neighboring stars would seem even more impossible to visit or send texts to.[116]

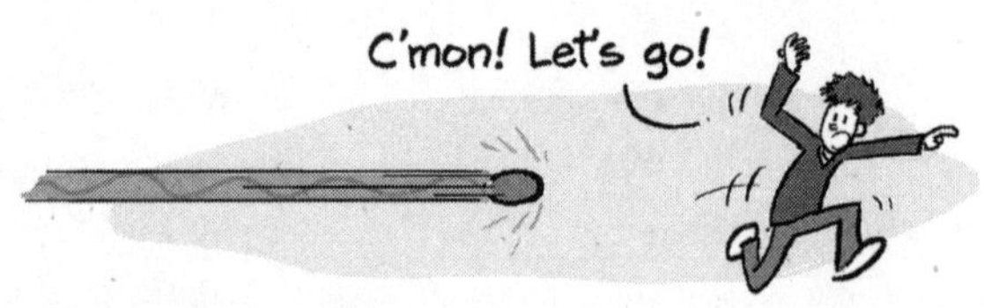

On the other hand, we can't blame it all on the speed of light. If space hadn't been stretched so much during the first fractions of a second after the Big Bang, everything would be much closer together today. And if dark energy wasn't currently pushing everything even farther away, then the prospects for interstellar travel wouldn't be getting worse by the minute.[117] We can imagine a universe where inflation had limited itself to blowing up the universe by a more reasonable amount than the absurd factor of 10^{25}.

So the emptiness of our universe comes from the interplay between these two quantities: the speed of light that defines the distance scales and the expansion of space, which is pulling everything apart. We don't know why either of these quantities are what they are, but if you changed them, you would get a universe that looks very different from ours. As with many of the big mysteries, we have only our single universe to study, so we don't know if this is the only way the universe could be organized, or if in other universes there was very little expansion and everyone feels much closer together than we do.

116 The roaming charges would be *killer*.

117 Uncool, dark energy. Very uncool.

Sizing It All Up

As you sip your hot caffeinated beverage of choice and look up at the night sky, reflect on the fact that everything we know about the size and structure of the universe comes from what we can see from Earth. Sure, we've sent probes to other planets, launched telescopes into space, and even put people on the moon, but from a cosmic perspective, we have basically gone nowhere. What we've learned about the universe we have surmised from looking up from our corner of the cosmos.

Despite this unassuming vantage point, we've been able to answer ancient questions (What are the stars? Why do they move?) and we have swept away long-standing misconceptions (that we're the center of the universe).

But how far out does it all go? Do we live in a finite or infinite universe? What will happen to the structure of the universe in a few billion years? The answers to these questions have enormous consequences for the panoramic view of ourselves and our place in the universe.

16. Is There a Theory of Everything?

What Is the Simplest Description of the Universe?

Only recently in human history has the world around us made much sense.

Before these past few centuries of scientific progress, it was a very common experience to be totally confused by everyday objects and events. What did early men and women think of lightning? Or stars? Or disease? Or magnetism? Or baboons? The world seemed to be full of mysterious things, powerful forces, and weird animals beyond our understanding.

ALSO MYSTERIOUS: BABOON MAGNETISM

More recently, this feeling has been replaced by a cool, casual confidence in science—the feeling that the world around us can be described by rational, discoverable laws. This experience is fairly new in the context of human history. It's not often that you encounter something completely mysterious in your everyday life. You almost never see things that shock you or that have no explanation. Lightning, stars, disease, magnetism, and even the mysterious baboons are largely explained as natural phenomena: things that are awe inspiring and beautiful but ultimately bound by physical laws. In fact, the experience of being at a loss for an explanation is so rare and novel that we *pay* to feel it again; it's what makes a magic show so much fun.

Beyond just understanding, we also have an impressively detailed mastery of our close surroundings; we can regularly fly four-hundred-ton airplanes across oceans, manage the quantum mechanics of billions of transistors in a computer chip, slice people open and insert bits from other bodies, and predict the mating habits of excited baboons. Truly, we live in an age of wonders.

But if we are so good at explaining the large trends and small details of our everyday world, does that mean we have it all figured out? Do our theories explain *Everything* (with a capital E)?

Unless you skipped the first chapters of this book, you have come to appreciate that the answer is a solid no. We are mostly clueless about what the universe is filled with (dark matter) and how to describe the most powerful forces that control it (dark energy, quantum gravity). It seems our mastery applies to only a tiny corner of the universe, and we are surrounded by a vast ocean of ignorance.

How do we reconcile these two ideas: that we understand the world around us but are mostly clueless about how the universe actually works? How close are we to discovering the ultimate theory: a *Theory of Everything* (ToE)? Does such a theory exist? Will it mean the end of all mystery in the universe?

It's time to go toe to toe with the universe's big ToE.

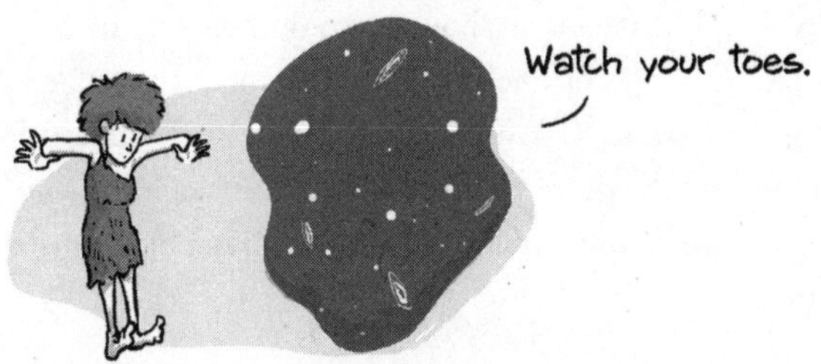

What Is a Theory of Everything?

Before we spend too much time talking about it, let's make sure we understand precisely what we mean by a "Theory of Everything." Simply put, a Theory of Everything would be the *simplest possible mathematical description of space and time and all the matter and forces in the universe at its deepest level.*

Let's break that down.

We include *matter* in the definition because this theory would have to describe everything that the universe is *made* of, and we include *forces* in the definition because we want this theory to describe more than just inert blobs of stuff. We want to know how that matter interacts and what it can do.

We also include *space* and *time* because we know that both concepts are malleable at some level and affect (and are affected by) the matter and forces in the universe.

Most important, we say *simplest* and *deepest level* because we want this theory to be the most fundamental description of the universe possible. Simplest means it should be nonreducible or bare bones (i.e., with as few variables or unexplained constants as possible). And *deepest level* means it should describe the universe at the smallest possible scale. We want to find the tiniest indivisible Lego blocks from which everything is made, and we want to know the absolute basic mechanisms they use to fit together.

TECHNICALLY, LEGOS
HAVE TWO TOES.

You see, we live in a universe that is like an onion. Not because it makes everyone cry when you slice it or because it's an essential ingredient in any great soup but because it's made up of layers and layers of *emergent phenomena*.

Take, for example, this model of the atom:

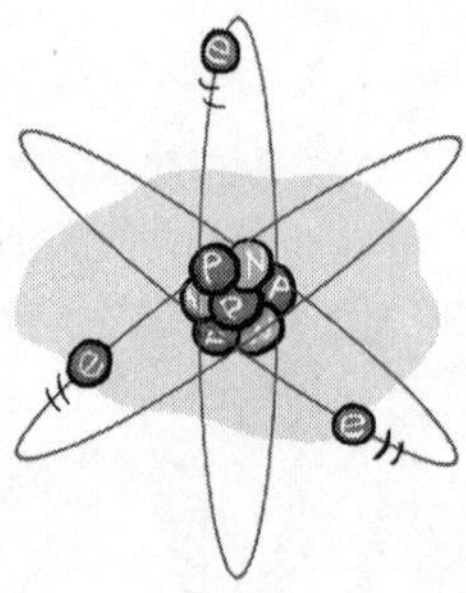

This diagram represents the theory that atoms are made of electrons orbiting around a nucleus made of protons and neutrons. It's probably one of the most recognized images in science. Coming up with it was an incredible achievement not just for the PR but because it meant that we moved beyond the idea that atoms are the fundamental units of matter to the deeper and more fundamental idea that they are made of even smaller bits.

But even that turned out to be an incomplete story. Some of these smaller parts are actually made of even smaller parts inside (protons and neutrons are made of quarks). On *top* of that, it turns out that things at this distance behave in a totally different way than we expected. In fact, it couldn't be any more different. Electrons, protons, and neutrons aren't little spherical balls with hard surfaces that clump together and swing around one another. They are fuzzy quantum particles defined by waves and governed by uncertainty and probability.

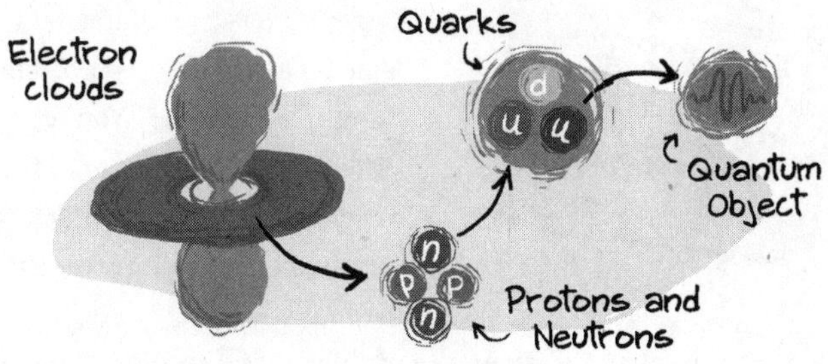

THE ATOM, REDUX

But all of those ideas work to some degree. The picture of atoms as little billiard balls works to describe how gas atoms bounce around inside a container. And the picture of atoms as little solid clumps with electrons swinging around them works to describe all the elements in the periodic table. And the new quantum view of particles works great to describe all sorts of natural phenomena.

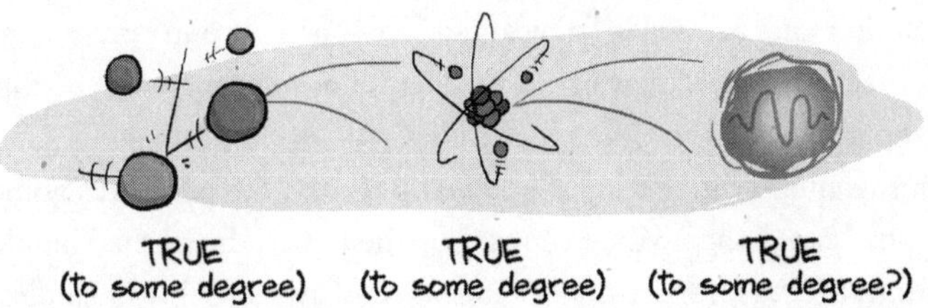

The point is that we seem to live in a universe where perfectly good theories can work even if they completely ignore what is happening underneath at smaller distances. In other words, you can accurately predict the collective action of all the little parts of something even if you don't know anything about what those little parts are doing (or if they even exist).

For example, the field of economics, which can mostly be described mathematically (assuming people resist their inner baboon and act rationally), is an emergent phenomenon of individual psychology. The actions of many individual shoppers and traders making buy or sell decisions cause large-scale changes in prices, which can be described using a few simple equations. You can study and describe the economics of a large group without understanding the choices and motivations of any individual.

And there are plenty of examples of this in physics. For example, even if we haven't discovered the most basic element of matter and still have no idea how gravity works as a quantum theory, we can still predict very accurately what will happen when a monkey jumps off the roof into your pool. We have very effective theories that can predict the projectile motion of the monkey; we have theories of fluid dynamics that can describe the resulting splash; and we have behavioral theories that explain why you do not like having a pool that smells of monkey.

In fact, there are layers and layers of these theories in the universe, each describing emergent phenomena at different levels. We had a theory of evolution long before we knew about DNA, and we put a person on the moon long before we knew about the Higgs boson or many of the fundamental particles we know and love today.

This is important because the *ultimate theory*, the one that's going to make physicists hang up their coats, drop their mics, throw their hands up in the air, say, "Yup, we're done," and walk away (probably unemployed) will be the one that describes nature at its most fundamental core. The ultimate theory will not describe some emergent phenomenon of the true building blocks of the universe; it will be *about* the true building blocks of the universe and how they fit together.

THE NOTORIOUS B.I.G. T.O.E.

This makes the concept of a Theory of Everything tricky because it might be that we'll never know with 100 percent certainty if we have reached that theory. It might be that we reach a theory that we *think* is fundamental but actually turns out to describe only the collective behavior of tiny submicroscopic baboons hidden underneath another layer of the universe onion. How would we know the difference?

Worse yet, what if the universe has an infinite number of layers? What if an ultimate theory is not even possible? What if it's baboons *all the way down*?

Baboons All the Way Down

Now that we've defined a Theory of Everything, let's explore the progress we've made in understanding nature at its deepest level regardless of whether it's necessary for getting monkeys out of your pool.

One question we can ask is whether there is a smallest distance in the universe. We are used to thinking about distances as having infinite resolution—that you can write a distance as 0.00000 . . . 00001 where the ". . ." represents an infinite number of zeros. But what if that's not the case? What if there is a distance below which smaller distances are not useful or sensible, like the pixels in your computer screen? If there is such a distance, then once our theory describes objects and interactions at that scale, we can be pretty confident that the theory is fundamental because there can't be anything smaller. But if there is no such distance, if things can be infinitely small or move infinitely small distances, then we may never be sure there isn't something else hiding underneath.

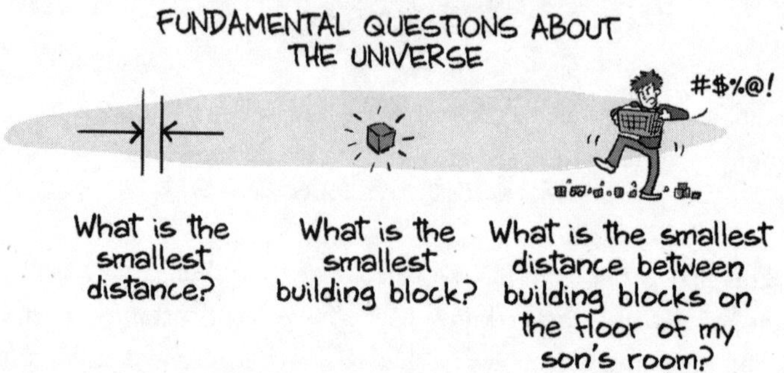

FUNDAMENTAL QUESTIONS ABOUT THE UNIVERSE
#$%@!
What is the smallest distance?
What is the smallest building block?
What is the smallest distance between building blocks on the floor of my son's room?

Another way to approach the question is to ask if the objects in our theory, the Lego building blocks it describes, are truly fundamental or if they are made of smaller Lego pieces. Are the electron and the quarks and the other particles we've found the smallest bits of matter in the universe? *Is* there even a smallest particle?

A final question is to ask how these objects interact: Are there different ways for them to interact (i.e., many different kinds of forces), or is there only one way to interact that manifests itself in different ways? What is the most fundamental description of the forces in the universe?

Let's start with the smallest distance.

The Smallest Distance

Is there a smallest distance, a fundamental resolution to our universe? Is reality *pixelated* at a scale below which it cannot be described? Let's take a moment to ponder how strange that is—the idea that reality could be pixelated.

Quantum mechanics tells us that we can't know with infinite precision the location of a particle. That's because in quantum mechanics objects

are actually fuzzy wavelike excitations of quantum fields with inherent random properties. But more than that, quantum mechanics tells us that the precise location of a particle is *not determined*, that information about its location below a certain distance doesn't exist. This is a clue that there might be some smallest meaningful distance to the universe, a quantization of distance that we could think of as pixelization.

But if reality is pixelated, how small are the pixels? We really have no idea, but physicists have made a very rough guess by looking around and combining several fundamental constants that tell us something basic about the universe. The first of these is the quantum mechanical constant, *h*, known as Planck's constant. This is a very important number because it is connected to the fundamental quantization of energy, which is like the pixelation of energy.

In order to arrive at a number that defines distances (for example, with units in meters), physicists multiply Planck's constant with two other constants: the maximum speed of the universe (*c*, the speed of light) and the strength of gravity (*G*). If we combine these in a particular way, we can come up with a number that has units of distance.[118] This number turns out to be very, *very* small: 10^{-35} meters, or 0.00000000000000000000000000000000001 meters.

We call this number the Planck length. What does this number mean? We don't really know, but it is *possible* that it gives us a rough estimate of the general size of the universe's pixels of space. There isn't really a justification for combining these numbers except that each of them represents a basic component of the physics that might be happening at the quantum

118 The Plank length is $(hG/c^3)^{1/2} = 1.616 \times 10^{-35}$m, where *h* is Planck's constant, *G* is the gravitational constant, and *c* is the speed of light.

level, so together they might give us a clue about the fundamental scale of the universe.

Can we confirm this? Not yet. Our tools for exploring tiny distances have advanced from optical microscopes, which can probe matter at the level of the wavelength of the light used (around 10^{-7} meters), to electron microscopes, which can probe matter at 10^{-10} meters. Beyond that, high-energy collisions in particle colliders have looked inside the proton at distances of around 10^{-20} meters.

Unfortunately, this means we are *15 orders of magnitude* away from examining reality at the Planck length. That means we are probably still missing *a lot* of detail. How much detail? Imagine if the smallest ruler you had or the smallest thing your eye could see was 1,000,000,000,000,000 (10^{15}) meters long. That's one hundred times bigger than the width of the solar system. If the smallest ruler you had was that long, there would be all sorts of amazing things happening that you would have no clue about. You can miss a lot of things in fifteen orders of magnitude.

Do we have any hope of exploring reality at the Planck length? Advances in technology have brought us from 10^{-7} (optical microscopes) to 10^{-20} (particle colliders) in a century or two, so it's not easy to project what future scientists will invent to give us even finer pictures of reality. But if we extrapolated from the strategy of using particle colliders, then seeing things at the Planck length would require accelerators with 10^{15} times more energy than the ones we have today. Unfortunately, such accelerators would need to be 10^{15} times bigger, which would cost 10^{15} times more, which is about 10^{15} more than what we can afford.

So we have no *solid proof* that there is pixelation at the smallest distances of the universe, but quantum mechanics and the universal constants we have measured so far strongly suggest that there might be, and that it's super-duper tiny.

The Smallest Particles

Are the electron, quarks, and other "fundamental" particles we've found the *most* fundamental particles in the universe? Probably not.

It seems fairly likely that the electron, the quarks, and all their cousins are actually just emergent phenomena of . . . something. Perhaps they are the result of a smaller, more fundamental particle or group of particles.

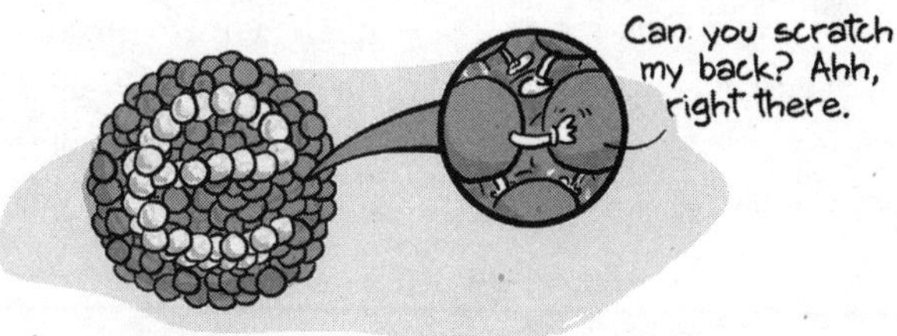

Can you scratch my back? Ahh, right there.

The reason we think so is that all the particles we've found so far seem to sit nicely in what looks a lot like a periodic table. Recall from chapter 4 that the smallest particles we've found so far can be arranged in a table that looks like this:

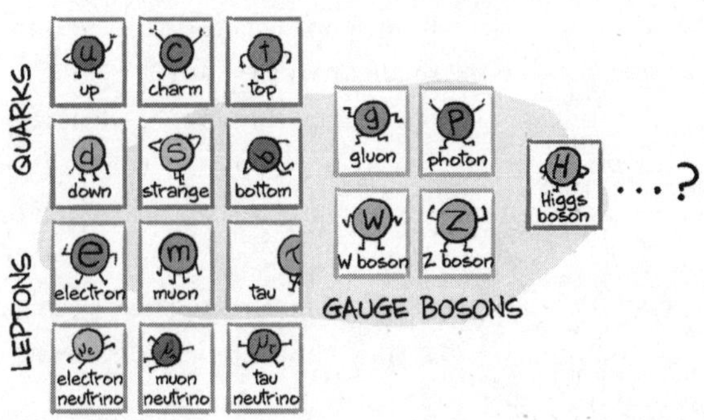

This neat arrangement and the patterns it seems to suggest tell us there might be something else going on here. Remember that the original periodic table of the elements (the one with oxygen, carbon, etc.) gave clues to scientists that all the elements are different configurations of electrons, protons, and neutrons. Similarly, this table makes physicists suspicious that the particles we've found might be made up of even smaller particles, or they might be the combination of one kind of smaller particle and some as-yet-determined law or rule that creates all the different variations of particles. In any case, the clues are there.

How will we know what is inside of electrons and quarks? We have to keep smashing things together.

If a particle is actually a composite particle (made up of smaller particles), then the smaller particles must be held together by some kind of bond with its own binding energy. For example, a hydrogen atom is actually a proton with an electron bound together by the electromagnetic attraction between them. In the same way, a proton is actually made of three quarks bound together by the strong force between the quarks.

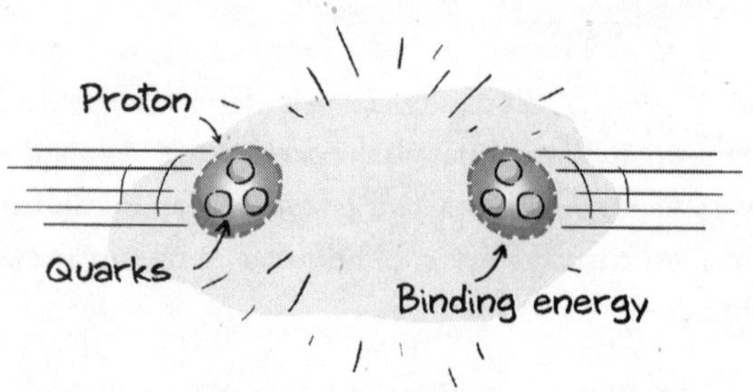

If you smash a composite particle with energy that is *less* than the energy of the bond between the smaller particles, then it's going to seem like a solid particle. For example, if a baboon throws a baseball very gently at your car, you would see the ball bounce off, and you and the baboon might conclude that your car is one giant single particle. But if the baboon throws the baseball *really* hard and the energy of the baseball is higher than the energy that's keeping all the car parts together, it might break a piece off, and you could tell that your car was pieced together from smaller bits, and probably made in America.

So one way to figure out if electrons and quarks are made of smaller particles is to smash them at higher and higher energies. If we reach a smashing energy that is higher than what might be holding the electron or quark together, then they would break apart and we would see that they are made of smaller pieces.

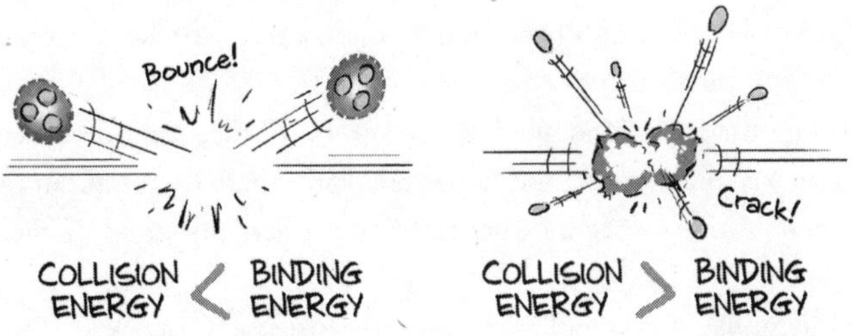
Bounce!
COLLISION ENERGY < BINDING ENERGY
Crack!
COLLISION ENERGY > BINDING ENERGY

But we don't actually know if electrons and quarks are made of smaller bits, and we have no idea what energy would be needed to break them apart if they *are* made of smaller bits. So far, our colliders, even the big expensive one in Geneva, haven't reached energies high enough to find any smaller parts to the electron, the quarks, or their cousins.

Another way we might figure out the patterns in the periodic table of fundamental particles is to find *new* particles that fit into the table. If we found *more* cousins of the electrons and the quarks, we might be able to deduce what the patterns in the table mean, which could lead to clues as to what their underlying structure is. This underlying structure might tell us whether there are smaller bits hidden inside our current set of particles.

WHAT DID YOU DO??
PHYSICS.

The Most Fundamental Forces

The final piece of building a Theory of Everything is a description of the fundamental forces in the universe.

We know that there are several different ways in which matter particles interact with one another, but how many forces are there? Could they all be part of the same phenomenon?

Finding the most fundamental description of forces in the universe is not about size (i.e., finding the "smallest" force there is); it's about finding out which of the forces we know about are actually parts of the *same thing*.

For example, if you had asked our prehistoric cave scientists Ook and Groog to list all the forces in the universe, they might have come up with a list that looks like this:

FORCES IN THE UNIVERSE:
By Ook and Groog

- The force that makes you fall off your llama.
- Whatever makes that shiny ball in the sky move.
- The force of wind.
- The force needed to break sticks.
- The force of a mastodon stepping on my toes.
- The force to get baboons out of the cave pool.
- Etc...

This list might contain many more seemingly unrelated experiences. But over the years, scientists have come to understand that many of these forces *are* related; many of them can be described by the same limited number of forces. For example, we know that the force that makes you fall off your llama is the same force that makes that shiny ball in the sky (the Sun) appear to move: it's gravity. And we know, for example, that the forces between objects (wind, sticks, mastodons) touching or pushing on one another are actually one force: the electromagnetic force between atoms that get close to one another.

In fact, the very idea that electric and magnetic forces are actually one force (electromagnetism) came relatively recently in the nineteenth century. James Maxwell noticed that electric currents make magnetic fields, and if you move magnets, you can make electric currents. So he wrote down all the known equations of electricity and magnetism (Ampère's law, Faraday's law, Gauss's law), and he realized that they have a perfect symmetry and could be rewritten in a way that treats electricity and magnetism as one single concept. They are not two distinct things; they are just two sides of the same coin.

More recently, this was done with the weak and electromagnetic forces. Two very different forces were found to also be two sides of the same coin: they could be written very simply as a single force (creatively called the "electroweak" force) with a similar kind of mathematical construction. The photon, which we all know and love, is actually just one feature of some deeper force that can also produce the W and Z bosons, which transmit the weak force.

Little by little, we have made progress to reduce Ook and Groog's long list of forces in the universe down to four forces and now only three.

FORCES	FORCE CARRIER PARTICLE
ELECTROWEAK	PHOTON, W AND Z BOSONS
STRONG FORCE	GLUON
GRAVITY	GRAVITON (THEORIZED)

How much further can we reduce the number of forces? Is it possible that all of these forces are actually part of the *same force*?

Is there just *one* force in the universe? We have no idea.

How Far Are We from a Theory of Everything?

A Theory of Everything needs to describe everything that exists in the universe in the simplest, most fundamental way possible. That means it must work down to the smallest distances of the universe (if such cosmic pixels exist); it must catalog the smallest Lego pieces in the universe; and it must describe all the possible interactions between the Lego pieces in the most unified way possible.

So far, we have some hints and some ideas about what the smallest distance in the universe might be (the Planck length). We have a pretty good catalog of twelve matter particles that so far we haven't been able to break further apart (the Standard Model). And we have a list of three possible ways that these particles can interact (the electroweak and strong forces and gravity).

How far are we from an ultimate Theory of Everything? We have no idea. But nothing can prevent us from speculating wildly.

If you follow the trends, then the simplest description of matter, forces, and space in the universe would presumably describe *one particle* and *one force,* and it would either describe the minimum resolution of space or confirm that there isn't one.

From this one theory, you should be able to trace *everything in the universe* (objects, behaviors, baboons) down through all the layers of emergent phenomena and explain them by the motions or behaviors of this one particle and one force.

So it seems like we still have a way to go. And lest we forget: all the theories we have so far only cover 5 percent of the universe! We still have no idea how we're going to expand what we know to the other 95 percent of the universe. Quite literally, we are barely tickling our ToEs.

Joining Gravity and Quantum Mechanics

One of the major obstacles for coming up with a Theory of Everything is joining gravity with quantum mechanics. Let's talk about that.

As it stands, we have two theories (theoretical frameworks, rather) for understanding the universe: quantum mechanics and general relativity. In quantum mechanics, everything in the universe, even forces, are quantum particles.[119] Quantum particles are tiny perturbations of reality that have wavelike properties that give them an inherent uncertainty. These perturbations move around in a fixed universe, and when they interact (when one of them pushes or pulls on another), they exchange other types of wavelike particles with one another. There are quantum theories for the strong force and the electroweak force, but there is no quantum theory for gravity.

General relativity, on the other hand, is a classical theory, which means it was invented *before* quantum mechanics. It doesn't assume that the world is quantized or even that matter and information are quantized. But one thing that general relativity *is* very good at is modeling gravity. In general relativity, gravity is not a force that two things with mass feel toward each other but a bending of space. When something has mass, it distorts the space and time around it in such a way that causes anything in the vicinity to curve toward that object.

119 A more modern and powerful description of quantum mechanics is quantum field theory, in which the basic elements of the universe are fields that exist everywhere and particles are places where the fields get excited, but that is beyond the scope of this book.

So we have a great theory for particles that covers most of the fundamental forces (quantum mechanics), and we have a great theory for gravity (general relativity), which is another of the fundamental forces. There's only one problem: these two theories are almost completely incompatible with each other.

It would be great if we could somehow merge the two theories because then we would have a common theoretical framework from which we could build a Theory of Everything. Unfortunately, that hasn't happened yet, and it's not for lack of trying.

When physicists try to merge quantum mechanics and general relativity, two big problems come up. First, there is the fact that quantum mechanics seems to work only on flat, boring, nonbendy space. If you try to make quantum mechanics work for gravity on curved, wobbly space, weird things start to happen.

You see, in order to make quantum mechanics work in the first place, physicists have to apply a special mathematical trick called renormalization. It's what allows quantum mechanics to deal with strange infinities, like the infinite charge density of a point particle electron or the infinite number of very low-energy photons that an electron can radiate. By using

renormalization, physicists can sweep all of these infinities under the rug and pretend there aren't any dead bodies hiding under there.

Unfortunately, when we try to apply renormalization to a quantum gravity theory with bendy space, it doesn't work. As soon as you get rid of one infinity, another one pops up. No matter how many of them you try to hide, there seems to be an infinite number of infinities. That means that all the theories of quantum gravity so far make crazy predictions that involve infinities, which means they can't be tested. The reason, as far as anyone understands it, is that gravity has a sort of feedback effect. The more that space curves, the more gravity there is, and the more that masses are attracted. So there's an apparent nonlinear feedback effect in gravity that you don't have in the quantum descriptions of the electroweak and strong forces.

The second problem with integrating general relativity into quantum mechanics is that both theories view the force of gravity so differently. If we were to incorporate gravity as a quantum mechanical force, then there has to be a quantum particle that transmits it, which nobody has ever seen. Technically, we haven't had the technology to detect such a particle (remember the graviton from chapter 6?) until recently, but so far, it hasn't been found.

So our two theories of how the universe works are hard to merge, and we don't even know if they *can* be merged. We have no idea what the graviton would look like, and all the predictions that a merged theory of quantum gravity makes skew toward infinity, which doesn't make any sense.

Either we don't have the right math to merge the two theories, or the way we are merging them is wrong. It's one of the two—or both! We know how to calculate forces in quantum mechanics, but we don't know how to use it to calculate the bending of space.

How Will We Know If We Are Done?

Let's imagine that scientists succeed one day in building a particle accelerator the size of the solar system (we'll call it the RLHC: Ridiculously Large Hadron Collider). And let's suppose that data from this absurdly high-energy collider reveals the fundamental element of matter at the Planck length, the smallest meaningful unit of distance.

Now let's further suppose that once we have this element of matter we are able to explain how this basic bit of matter interacts with itself and comes together to form the emergent phenomena of nature at the larger distances.

Would that mean we're finished?

Ever since William of Ockham,[120] scientists and philosophers have preferred simpler, more compact explanations over longer, more complex ones. For example, suppose you came home one day and your pool smelled like baboons. Would it make more sense to assume that an international crime organization put drops of baboon perfume in your pool as part of a complicated heist involving Justin Bieber and three professional basketball players, or would it make more sense to simply assume your pet baboon disobeyed your order and jumped in the pool to cool off?

If you have two competing theories that both explain the data, the simpler one is more likely to be correct (assuming you own a baboon). Physicists have had very good luck at successively simplifying our theories by noticing when different phenomena are actually complementary sides of the same coin, like thunder and lightning.

But similar to the question of "Is there a smallest particle?" we can ask, "Is there a simplest theory?" We might be able to prove that the universe has a smallest distance, and maybe a tiniest particle, but could we *prove* that we have the simplest theory? How will we know when we are finished? We might think we have achieved it and then meet an alien race whose physicists have an even simpler theory.

120 In the fourteenth century, William invented Ockham's razor (also called Occam's razor), a breakthrough in shaving technology and the first expression of the idea that simpler explanations should be preferred.

The first thing to consider is how we measure the simplicity of a theory. Is simplicity measured by how compactly you can write the theory down? By how beautifully symmetric the equations are? Whether it will fit on a T-shirt?

One important criterion is the number of numbers in it. For example, let's suppose you come up with a theory of everything, and in your formula, there is a number. It doesn't really matter what the value of the number is, but let's say that it's important, like the mass of the most fundamental particle, the "tinyon." And let's suppose that in order for you to use the theory (say to predict how long the fall from a llama will take), you have to know the value for that number. Naturally, you would go back to your collider and use it to measure the mass of the tinyon, which you then go and plug into your theory. Voilà, your theory is done, and you sit in your dented car to wait for the Nobel Prize committee to announce your imminent award.

But now suppose someone else comes along, and she says *she* also has a Theory of Everything. But *her* theory does something different: it comes built-in with the exact value of the mass of the tinyon, and it will not work unless it has that one precise value. She doesn't have to go out and measure it, her equation *tells* her what it should be. It has one less arbitrary variable than yours.

Although your equation might seem more general than the other person's, hers actually tells us more about the universe. This is because her equation would tell us *why* the mass of the tinyon has to be what it is (otherwise the theory wouldn't work). It has fewer numbers, which means it's simpler, which means it's more fundamental. Good-bye, Nobel Prize.

The point is that one way to know if we have reached the ultimate

Theory of Everything is by counting how many arbitrary numbers it has. The fewer the numbers, the closer we must be to the center of the onion.

Perhaps there *are* no numbers at the center. Maybe at the core of this bulbous root of a universe there is only fancy math, and all the numbers we know about (such as the gravitational constant or the Planck length or the number of times mastodons have stepped on your feet) can be beautifully derived by this mathematics.

Currently, the Standard Model has many such parameters—twenty-one of them are described in the following list—and it doesn't even pretend to describe gravity, dark matter, or dark energy:

Twelve parameters for the masses of quarks and leptons
Four mixing angles that determine how quarks change into each other[121]
Three parameters that determine the strength of the electroweak and strong forces
Two parameters for the Higgs theory
One partridge in a pear tree (theorized)

The truth is that we have no idea how to determine if a theory is the final theory. It could be that there are no arbitrary numbers to the universe. Or maybe there are and they have deep meaning. If we discover

121 Recent discoveries that neutrinos can also change into each other mean four more parameters.

what seems to be a final theory, and it has the number four in it, does that mean that there is something deeply important about the number four?

Or perhaps such basic numbers could have been set at random during the early moments of our universe and that in other pocket universes they have different values. See chapter 14 for a discussion of such multiverses, but be warned that most of these ideas depart from falsifiable scientific hypotheses and dive deep into untestable philosophical theories.

Reaching for Our ToEs

Since we are fifteen orders of magnitude from probing the Planck length, and we're still struggling to find a unified theory that can describe a meager 5 percent of the universe, perhaps it's time to try an alternate approach.

What if, instead of drilling down through the layers of the onion, we start at the center?

Right now, we are so far from the center of the onion that we can freely speculate about what reality looks like there.

Perhaps the universe is made out of one kind of tiny particle or little cocktail wieners or miniature baboons.

As long as your hypothetical ToE eventually predicts the particles and forces we see today, there is technically nothing to contradict your theory. If that makes it sound like the nature of the universe is just a vast intellectual playground with no rules, that is correct, but only if you are a philosopher or mathematician. If you want to be scientific about it (ahem, physicists), your miniature-baboon theory has to do more than describe how electrons are made out of "baboonitos." It also has to make some sort of testable prediction so that we can verify it and distinguish it from theories about tinyons and wienerons.

String Theory

Probably the most popular and controversial approach in modern theoretical physics is string theory, which suggests that the universe has ten or eleven dimensions of space-time if not more. Many of these new dimensions are not visible to us because they are rolled up or very small (see chapter 9 for more discussion of how this is not totally made-up gibberish), and they are filled with tiny little strings.

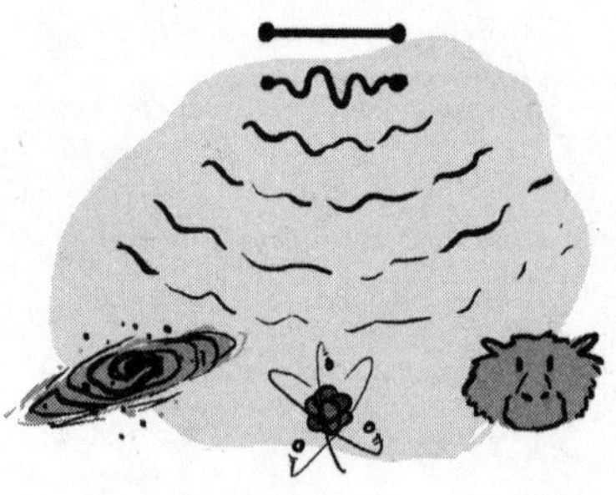

These strings vibrate and, depending on how they are vibrating, can look like any of the particles we have already discovered. They can even describe particles we haven't seen yet, such as the graviton. Even better, string theory is supposed to be mathematically beautiful and theoretically fascinating. String theory is a true ToE because it unifies all of the forces and describes reality at its most basic level. Before you sign your name to the list of true believers at the Church of String Theory, you should be aware of a few little details. Or we can call them issues. Okay, concerns. Well, maybe they are big problems.

The first problem is that while string theory *can* describe the whole universe it hasn't yet done so. So far, physicists haven't found a reason why string theory *can't* be a ToE, but the theory is still far from complete.

The mathematics are still being developed, and some pieces still need to fall into place before it can be considered a complete descriptive theory.

And that brings us to the second problem: string theory is still only a *descriptive* theory and can't yet make any predictions that we can test. Just because a theory is fully consistent and mathematically attractive doesn't mean it's a scientifically valid hypothesis.

In order to know whether the smallest bits of the universe are little tinyons or vibrating strings, each theory has to make a testable prediction. Since string theory deals with objects only at the Planck length so far, it can't be scientifically tested yet. Just as with the Deep Space Kitten theory, it may or may not be true, but believing in it without experimental verification is a question of philosophy, mathematics, or faith, not physics.

STRING THEORY IS ACTUALLY CAT WHISKER THEORY.

It is certainly possible that one day in the future experimental techniques will vastly improve or that clever theorists will find a feature of the universe at testable distances that is a unique prediction (and therefore a test) of string theory. But not yet.

The final problem with string theory is one of parameters. The dynamics predicted by string theory are determined by the number and shape of space-time dimensions. And there are a lot of ways one could choose these dimensions. More than a lot—something like 10^{500}, which is 10^{410} more than the number of particles in the universe and 10^{497} more than the number of friends you have on Facebook. There is hope that new formulations of string theory will reduce the number of arbitrary choices, but if you want to judge a theory's completeness by its number of parameters, this one still has a long way to go.

Getting Loopy

A completely different approach imagines that at the smallest level space is quantized. In this theory, space is built out of tiny indivisible units called loops that are the size of the Plank length, or 10^{-35} meters. If you weave enough of these loops together, it might be possible to derive all of space and matter.

This theory, called loop quantum gravity, can unify gravity with the other forces and explain the nature of the universe down to the smallest bits. Unfortunately, it suffers from the same difficulty as string theory: without a way to verify it, it can't be promoted to a scientific theory. There *is* one specific prediction that it makes, which is that the Big Bang was part of a cycle called the Big Bounce, in which the universe repeatedly expands and contracts. But while it may be possible to validate this theory, you'd have wait billions of years for the next Big Bounce to happen before you can claim that coveted Nobel Prize.[122]

THE RAINBOW LOOM THEORY OF EVERYTHING

These are just the first few tentative steps. On top of these ideas, or inspired by these ideas, or out of a crazy meditation session while surrounded by thoughtful baboons, we hope that some physicist will build a Theory of Everything that does explain everything and makes testable predictions.

122 Not awarded posthumously, so if you die in proving your theory, it's a double bummer.

Would It Even Be Useful?

How useful would a Theory of Everything be for answering questions about everyday objects?

In practice, not very.

Even though a Theory of Everything would reveal to us the inner workings of the universe at its most fundamental level, it will probably not be very useful for practical things, such as designing a monkey-proof net to cover your pool.

What's interesting about the idea of the universe as a multilayered onion of emergent phenomena is that different theories at different levels can all be correct *at the same time*. For example, suppose you want to describe the motion of a bouncing ball. You can describe that using Newtonian physics (the kind you learned in high school) and treat the whole ball as one object being pulled on by the force of gravity. In this case, you would get a simple parabolic motion that can be written in a single line of math.

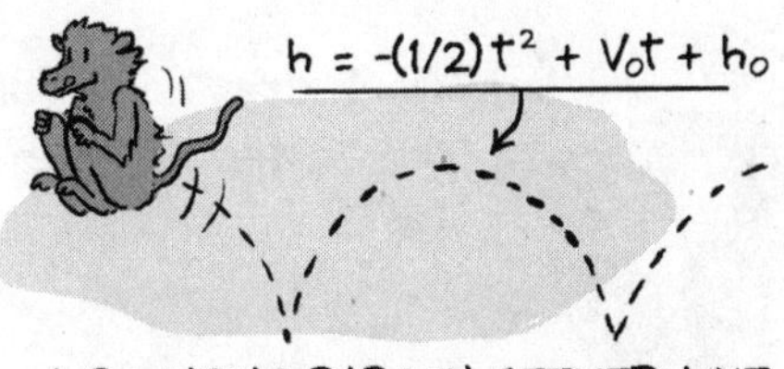

A BOUNCING BABOON SEEMED LIKE IT WOULD MAKE A BETTER GRAPHIC.

Alternately, you could also describe the bouncing ball using quantum-field theory. You could model the quantum mechanics of every single one of the 10^{25} or so particles inside the ball and track what happens to each and every one as they interact with one another and the environment. *Totally impractical*, but possible in principle. In theory, that should give you the same result as above, but in practice, it's almost impossible to do.

If we had a correct theory of the lowest level of reality, we could *in*

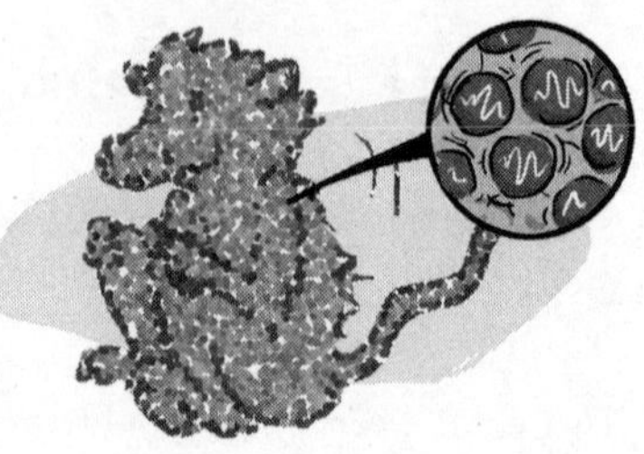

principle derive the formation of galaxies or fluid mechanics or organic chemistry from that theory. But practically, it would be ridiculous, and it's not a useful way to do science.

Amazingly enough, the universe is understandable and describable at multiple levels. You don't have to start from the lowest level to do organic chemistry or understand our baboon obsession. It would be a huge pain if you did, right? Nobody expects a surfer to understand string theory and compute the motion of 10^{30} particles in a wave in order to stand up on her surfboard. Similarly, when you are baking a cake, you wouldn't want to get the recipe in terms of quarks and electrons.[123]

Step 1: Create Big Bang
Step 2: Wait 14 billion years
Step 3: ...

NOBODY WANTS AN ACCURATE COOKBOOK.

If early scientists had had to start from the very basic particles when humanity began our journey of discovery, then we wouldn't have gotten anywhere.

123 Your local supermarket sells plenty of quarks and electrons, but they are not individually packaged.

From Head to ToE

The quest to find a Theory of Everything is an attempt to do something we have never before accomplished in science: reveal the deepest, most basic truth of our universe.

So far, we have proven ourselves to be pretty good at building useful descriptions of the world around us. From chemistry to economics to monkey psychology, we've put a lot these descriptions to work improving our lives and helping us build societies, cure diseases, and get faster Internet speeds. That these descriptions are not fundamental and describe only emergent phenomena doesn't make them any less useful or effective.

But one thing that these theories are missing is the satisfaction of revealing how the universe *really* works.

And we want to know the deepest truth. Not because it would help us solve our baboon's behavior problems or improve our Netflix binge-watching, but because it would help us understand our place in the universe.

Unfortunately, as with most of the big questions in the universe, we have no idea what a Theory of Everything would look like. Right now, we suspect that the tiniest particles we know about (electrons, quarks, etc.) may be *10^{15} times* larger than the basic building blocks of the universe. Imagine being the size of a galaxy and thinking that a star is the smallest thing in the universe. That's how far off we might be from a true Theory of Everything.

And we still haven't succeeded in describing all of the forces in terms of a single theory. Gravity still doesn't play nicely with quantum mechanics despite a century of mediation and pet therapy. We don't even have a guarantee that there *is* a Theory of Everything in the universe.

But none of that should dissuade us from looking. So far, every time we pull back a layer of reality, every time we take a step toward the core of the universe onion, new and bizarre structures are revealed that make us think differently about the way we live our lives.

WARNING: PHYSICS MAY GIVE YOU ONION BREATH.

17.
Are We Alone in the Universe?

Why Has Nobody Come Around to Visit?

If you travel to a foreign country, you will make the charming discovery that there are many differences between the local way of life and your own.

Is a cup of coffee over there huge and watery or tiny and head-poppingly strong? Do the bathrooms have little rooms with doors that close for privacy or just flimsy stalls that do nothing to hide your traveler's indigestion? Does nodding your head mean yes, no, or that you want extra eyeballs and tentacles in your smoothie? Do they eat with forks or sticks or their hands or use trained butterflies? Do they drive on the left side, the right side, or every side?[124] Even more important, do they

STRANGER IN A STRANGE STALL

124 We're looking at you, Italy.

organize their lives around accumulating money or finding love or annoying their relatives?

On the other hand, you will also find that many things are similar to your lifestyle at home: over there, they still eat, sleep, and talk to one another. Maybe their breakfast has little eyeballs staring back at them, or they drink weak coffee served out of a shoe, but in the end, they eat and drink just like you do.

The important point is that visiting other cultures lets you learn which parts of your culture are *universal* to humans because they come from deep fundamental needs of humanity (eating, sleeping, caffeine, etc.) and which are *local* random choices that might seem fundamental to us (toilet stalls, utensils, tentacles for breakfast, etc.) but could easily have been different. Seeing another culture is the best way to learn which things you thought were universal but are actually local.

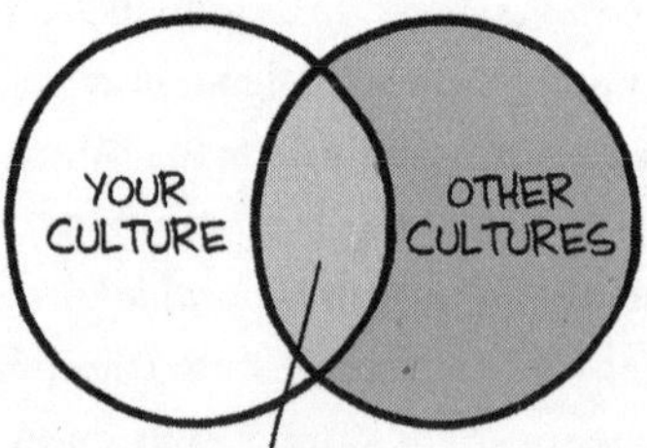

The same principle that applies to breakfast foods is also true for science. Many of our misconceptions about the universe come from overgeneralizing our tiny local experience. For example, for millennia humans imagined that we were at the center of the universe or, even worse, that our world *was* the whole universe, with the stars and the Sun as props made just for us. These were totally reasonable ideas given our local experience.

Maybe in five thousand years we will look back on our current views as embarrassingly naïve. Astronomy has already taught us the difficult lesson that we are just tiny people on a tiny speck in a not very special corner of a ginormous universe. What else have we misunderstood because we see the universe from only this single vantage point? What about the universe do we assume to be universal when it is really just local? Can you even get good boiled eyeballs to go at three a.m. around Alpha Centauri?

But perhaps the most important question we can ask about the universality of our experience is the question about life itself: is life in the universe common or rare?

Is the universe teeming with life, or are we the only ones out here? Having only explored the Earth and our immediate neighborhood, it's hard to draw conclusions about whether or not we are alone in the universe. Are we like some primitive uncontacted tribe living in the middle of the jungle that is clueless about the vast civilization around us? Or are we more like an isolated oasis of life in a huge, empty, sterile desert? Unfortunately, both possibilities match our local experience so we can't tell the difference.

If there is intelligent life out there—a huge if—the second question to ask is: Why have we never met them? Why haven't we received any messages, letters, or birthday party invitations? Are we the only ones awake in the universe, or are the other civilizations just too far away or ignoring us on purpose to leave us as cosmic outcasts in the galactic game of dodgeball?

Finally, if we were contacted by intelligent technological life, what could we learn from talking to them? What have they figured out about the world that we haven't? We have mostly explored the universe using electromagnetic radiation (i.e., light) because that's what we use to explore with our eyes. Maybe these aliens discovered that the universe is bathed in some other form of information (neutrinos or some particle we don't know about yet) and have a completely different picture of how everything works. Maybe they don't even have eyes! This is all wild speculation, but all of these scenarios are possible, and we have no idea which scenario corresponds to our universe.

Even the idea of learning from aliens makes a lot of assumptions about how sentient life conducts its business. Do they write books or just zap the information to one another via direct brain connections? Is math part of their thinking or is it a human invention? Do they even do science? We did zero science until embarrassingly recently. Even now, our science is mostly

just drinking coffee with the occasional flash of insight and rare afternoon of actual progress.

In this chapter, we will discuss our state of knowledge and ignorance surrounding some of the deepest questions about life itself: Are we alone? If we are not alone, why hasn't anyone contacted us? Do we want to contact *them*? If they contacted us, what could we learn about life, the universe, and everything?[125]

Are They Out There?

If we are the only life in the entire universe, then there is something very, very strange about our experience and our planet. Being alone in such a vast cosmos would mean that life is extremely rare. If the universe is infinite, then being the sole example of something is more than rare, it is practically impossible. In an infinite universe, anything with even a tiny probability occurs. In fact, anything with a finite probability occurs infinitely often. Only things with infinitely small probabilities occur exactly once.

On the other hand, if we are not alone, then it cements the feeling that life and perhaps even our intelligence and civilization don't give us a special place in the universe. It would mean that very little about the human experience reveals anything deep or interesting about the universe itself. That is both humbling and exciting.

Which is it? Are we special or boring?

The problem is that it's very difficult to extrapolate from our single-planet experience to a more general understanding. There are two possi-

125 Besides "42."

We're special!
That's boring.

bilities, and we can't distinguish between them: either (1) we are the only life in the universe, or (2) the universe is teeming with life that we haven't been able to see either because it's too far away or too alien for us to notice or recognize.

Imagine you are an elementary school student. And one day your normal math quiz unexpectedly comes with an answer sheet! At first you are excited, but then you start to wonder: Are you the only one who got the answer sheet? Maybe this is a practice exam, and nobody told you. Or maybe there are other kids who got the answer sheet, but they don't want anyone to know they have it. You have no idea if you are the only student to be so fortunate or if everyone has the answers. If none of the other students have the answer sheet, they would never know to ask. The fact that you have it tells you nothing about whether you are special or not. You can't know everything about the bigger picture from your local experience.

In the case of life, we can do a little bit better than that but not much. For example, we can look around on Earth and study the various forms of life that do exist. If there are features that vary widely from organism to organism (e.g., skin color, favorite ice cream flavor) then we can be confident that they are not essential or fundamental to life and that life on other planets might be totally different (maybe garlic ice cream is a huge hit on the planet Zlybroxxia). On the other hand, if there are things that are constant to all forms of life on Earth (e.g., the need for an energy source and water), we can speculate that they might be common to life everywhere. This argument is especially strong because we can show that common elements of life have developed several times independently—eyeballs, for instance (no joke!).

It can be helpful to separate some of these issues by writing the question down as a math problem. For example, if you wanted to estimate the number of people in your neighborhood, you could do it by performing an exhaustive door-to-door survey, *or* you could do it by simply multiplying the number of houses in your neighborhood by the average number of people in a typical house.

Similarly, we can estimate the number of intelligent species we could potentially talk to (N) as a math equation that looks like this:

$$N = n_{stars} \times n_{planets} \times f_{livable} \times f_{life} \times f_{intelligent} \times f_{civ} \times L$$

Where the pieces are:

n_{stars}: The number of stars in the galaxy

$n_{planets}$: The average number of planets per star

$f_{livable}$: The fraction of those planets that could support life

f_{life}: The fraction of livable planets that actually develop life

$f_{intelligent}$: The fraction of planets with life that develop intelligent life

f_{civ}: The fraction of intelligent species that develop technological civilization and can send messages or spaceships into space

L: The probability that they are around at the same time as we are

This is a very simple mathematical formulation (known as the Drake equation), but it's useful because it breaks the problem into parts and shows that if just *one* of these pieces is zero then we will never hear from aliens even if they do exist.

But keep in mind that this is just an estimate based on our local experience of life. In the end, we are fundamentally limited by our lack of interplanetary tourism. We might draw up a careful list of the most general requirements for life, but they might just be for life as we know it. It's entirely possible that life could take forms that we can't currently imagine, with metabolisms that run incredibly slowly and life cycles that seem impossibly long or organisms that are absurdly vast or whose boundaries from their environments or one another are fuzzy. So keep in mind that we could be totally wrong about the requirements for intelligent life and that the only way to know for sure is to find examples in other parts of the universe.

With that caveat, let's tackle the pieces of this equation one at a time.

Number of Stars (n_{stars})

Astronomers have determined that there are a huge number of stars in our galaxy: 100 billion. It feels encouraging to start from such a big number because the rest of the pieces in the equation could all be tiny probabilities.

But why focus on just our galaxy? There are an estimated one to two trillion other galaxies in our observable universe. The reason we start with the Milky Way is that while the stars in our galaxy are very far away other galaxies are depressingly distant. And travel or communication at those scales seems nearly hopeless unless we rely on loopholes such as wormholes or warp drives. Let's focus on our galaxy for now and keep the extra factor of a few trillion other galaxies in our back pocket to boost the numbers if we get too discouraged.

The book said to keep
a few trillion galaxies
in my pocket.

Number of Planets Suitable for Life
($n_{planets} \times f_{livable}$)

Of all the stars in our galaxy, how many have planets that can harbor life? And what kind of planet can harbor life? Are rocky planets like the Earth the only type, or are there many possible homes for life?

For example, maybe there are forms of life that can live high atop the atmospheres of enormous frozen gas giants, or maybe there are life forms that can swim in lava on the molten surface of tiny hot planets.

For now, let's focus our search on the number of Earthlike planets, which means rocky worlds rather than gas planets, and those of a similar size and with a similar amount of solar energy. It's more limiting to think this way, but it's also more realistic, given that Earth is the only planet we know of that has life.

So how many cozy planets like ours are there in our galaxy? Our telescopes are not powerful enough to see the tiny little dark rocks that might be orbiting distant bright stars. Not only are those planets so far away that they are essentially invisible to us, but they are much closer to their stars than they are to us, which means they are hopelessly outshined. If you are staring into a huge bright spotlight, you will never notice a tiny little speck of rock right next to it.

This is why we had no idea until recently how many planets were

around a typical star and how many of them were similar to the Earth. But in the past decade or so, astronomers have developed some very clever techniques for indirectly detecting planets. They can look for a small wiggle in a star's position, which means that the star is being pulled slightly by the gravitational force of a planet. They can also look for periodic dips in the light from the star, which means that the planet orbiting the star is passing in front of it. Using these techniques and others, astronomers have discovered something incredible: about *one in five* stars in our galaxy has a rocky planet with a size similar to Earth's and a similar amount of solar energy on its surface. That means that the number of possible Earths just in our galaxy is in the *billions*. Woo-hoo! Good news so far for the nascent alien tourism industry.

1 OUT OF 5 STARS

Proportion of stars with an
Earthlike planet in orbit
OR
Average review for that new
eyeball-smoothie restaurant

Number of Habitable Planets with Life (f_{life})

If we focus just on our local galaxy, we know that there are about 100 billion stars, with about 20 billion Earth-like planets. Twenty billion makes for a lot of petri dishes for creating life. So the numbers seem encouraging, but now we get into more difficult waters: How many livable planets actually *have* life?

To start thinking about this, we can first ask: What are the necessary ingredients for life? From studying the huge variety of life on the Earth, we can conclude that it always seems to require water in order to do a lot of the complex chemistry and transport, and it also seems to require plenty of carbon in order to make many of the complex chemicals and provide structural support, such as cell walls and bones. In addition, it tends to require nitrogen, phosphorus, and sulfur, mostly in order to make DNA and critical proteins.

Can life as we know it form without these elements? Some have speculated that silicon could serve in place of carbon. That is a fun example of trying to think broadly, but since silicon is much heavier and more complicated (fourteen protons) than carbon (six protons), it's probably not abundant enough to open up many new paths for life.

A trickier question is whether these ingredients are sufficient for life. If you had a nice warm planet somewhere with huge oceans and plenty of these elements sloshing around and banging into one another, what are the chances that life will begin? This is one of the deepest and most basic questions in biology, but one that is very hard to answer. Here on Earth we know that life began a few hundred million years after there was water on the surface. But we know little about the details, and we certainly don't know if this is an unusually short or long time to have to stir the chemical soup and wait.

Scientists have tried to replicate some of the steps thought to be needed to go from sterile soup to living organisms. A famous experiment started with a soup of these chemicals and added an electrical spark to mimic the effect of lightning strikes in a primitive Earth. No Frankenstein was created, but a few of the complex molecules that are needed for life were formed. This suggests that—for some of the steps, at least—maybe you only need to have the pieces lying around and then wait for the right injection of energy through geothermal heat, lightning strikes, or alien laser weapons.

So we understand very little so far about how life was created from a sterile environment on Earth.[126] If we knew more, we could make some reasonable argument about the likelihood of life as we know it getting started on other planets with similar conditions. Until then, we simply have no idea if a setup like ours gives life every time or only once in a million or quadrillion times. And, remember, there could be other drastically different forms of life, each with their own probability of getting started from a sterile soup.

It turns out that the Earth is not the only place in our neighborhood where the chemical building blocks of life exist. Many have been found on Mars (including liquid water!), but so far, there is no evidence of life, large or small.

Other places in our solar system might not make your top five vacation destinations, but they are reasonable candidates for hosting life. Jupiter's moon Europa is thought to have a huge underground ocean, and Saturn's moon Titan has an atmosphere and oceans of chemicals that could be used to build early life-forms. This is a far cry from finding actual life out there, but at least the ingredients seem widespread.

126 Especially these authors, neither of whom are biologists, but even biologists we know admit to similar ignorance.

While we speculate without basis, how certain are we that Earth is where life began? Of all the implausible possibilities, one that sounds like science fiction but has some nonzero chance of being true, is that *life began elsewhere* and traveled to Earth via meteorites.

We hear you scoffing at this idea, probably because you are imagining microbes building microrockets and taking zillion-year-long journeys to land on Earth. Actually, microbes don't need to build their own rockets to move between planets or stars. When something big (like a huge asteroid) hits a planet, the impact can send bits of the planet out into space. Those bits fly around for a while—sometimes a very long while. Sometimes they drift in space for billions of years, and sometimes they are fried by passing too close to a star. But occasionally they fall to another planet. Scientists have found rocks on Earth that almost certainly came from Mars via this mechanism. So it is certainly possible for rocks to get blasted from one planet to another. If those rocks happen to contain living organisms sheltered on the inside, or tiny microbes and even micro-animals that can survive the vacuum of space[127] then it is not impossible (though still somewhat implausible) that microbial life could hop from planet to planet.

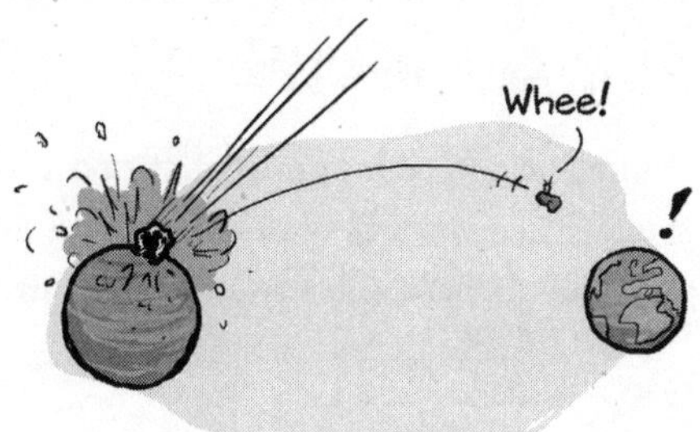

If this was true—and we have zero evidence to support it—it would mean that aliens do exist: *we* are the aliens! In fact, scientists once found a rock that clearly came from Mars and even had strange unexplained lifelike shapes inside of it. Those shapes very roughly resembled microbial life on Earth, but many scientists are deeply skeptical that they are evidence for Martian life. However, it does prove that if there was life on

127 Google the word "tardigrade" and prepare to be shocked.

Mars (or elsewhere) it could have hitched a ride to a young Earth and seeded it with life.

Other than speculating about whether our great-great-great-great-grandparents were extraterrestrials, this idea gives us an opportunity. If life exists on other planets, we might be able to discover evidence of it by examining asteroids. These pieces of interplanetary junk might not have the conditions to create life, but if they were blasted off of a planet far away, they could carry evidence of life from those distant worlds.

Number of Habitable Planets with Intelligent Life ($f_{intelligent}$)

Once you get microbial life going, what other conditions do you need in order to form complex life and then intelligent life?

You certainly need enough time, which means you have to have long periods between events that could destroy a fragile initial colony. On Earth, intelligent life appeared fifty thousand to one million years ago, depending on your threshold for intelligence (some would argue we haven't reached it yet). That is billions of years after life began, so it's not a rapid process.

This puts some constraints on where life might be possible. For example, if your planet is too close to the center of the galaxy, it will be bathed in punishing radiation from the central black hole and neutron stars. This radiation could decimate the delicate chemistry of life.

USEFUL INGREDIENTS FOR LIFE:

✓ CARBON
✓ WATER
✓ PHOSPHORUS
✓ NITROGEN
✓ SULFUR
✓ SUNBLOCK

There is another good reason you don't want to be too close to the older stars and the dense galactic center: all of those nearby objects can bump or gravitationally perturb the orbits of large meteors and asteroids in your solar system, causing extinction events when some of them smack down on the surface of your planet. In our solar system, some scientists speculate that having two massive planets (Saturn and Jupiter) with orbits farther than ours from the Sun operates as a sort of cosmic vacuum cleaner, picking up many objects that might otherwise be a danger to Earth.

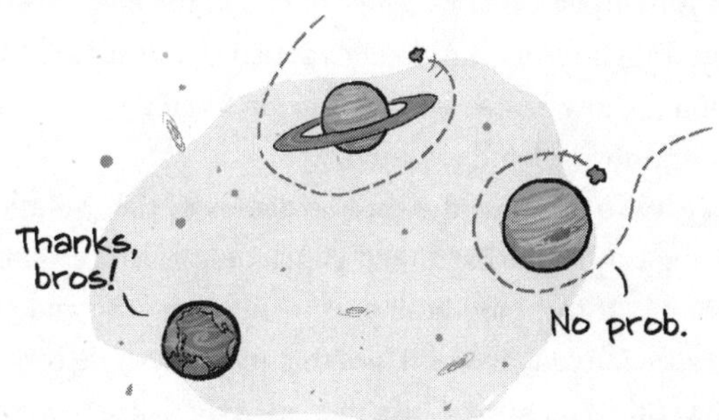

On the other hand, you can't be too far from the center of the galaxy because you need to have enough of the heavier elements to make complex chemistry. These elements can be formed only through fusion in the center of stars and then dispersed when those stars collapse and explode. These stars are rarer near the edges of the galaxy, so the planet can't be too far from the center. But you need more than just enough time; perhaps intelligent life is not inevitable and requires good luck or special circumstances. Is it necessary for intelligence to have dexterous hands in order to develop tools and manipulate the environment? Does technological civilization require complex social groups in order to form language and symbolic thought? If the dinosaurs had not been wiped out by that huge asteroid, would intelligent life exist on Earth today or ever? We have no idea.

In short, we have almost no information on how often life turns into complex life or develops intelligence or technology. Many people speculate on these questions, some even making reasonable-sounding arguments

for why life should be rare or common. But in the end, most of these arguments extrapolate from our local experience and suffer from the same flaw: we don't know which aspects of our intelligent life are local and unessential or universal and essential.

It is too easy to examine the specific details of the evolution of intelligent technological life on Earth and conclude that all of these details are necessary. Some of them are bound to be idiosyncratic and perhaps even vanishingly rare in the universe. Does that mean that life is rare? Not necessarily. The critical question is whether our experience represents the only possible path to life as we know it, one of many possible paths to life as we know it, or one of many possible paths to life as we never imagined it.

This factor, $f_{intelligent}$, could be 1, or 0.1, or 0.0000000000001, or smaller.

Number of Civilizations with Advanced Communications Technology (f_{civ})

For the sake of argument, let's pretend that the parts we have considered so far ($n_{stars} \times n_{planets} \times f_{livable} \times f_{life} \times f_{intelligent}$) still give us a large number of intelligent species in our galaxy. We have no good reason to pretend that this is true other than it lets us continue thinking about the other bits and avoids an abrupt end to this chapter.

If there was other intelligent life in the galaxy, even living around nearby stars, how could we detect it? We explore the universe mostly using the broad spectrum of electromagnetic (EM) radiation: radio waves, visible light, X-rays, etc. Our preference for using EM radiation is rooted in our love of vision because that's what our eyeballs use. But what do the aliens use? Perhaps they prefer to send messages with beams of neutrinos or shockwaves in dark matter or ripples in space itself. We have no idea what their primary sensory organs might be (or if they have sensory organs) and what they might be sensitive to.

Another possibility entirely is that they don't communicate via radiation but instead send robotic probes to explore the galaxy. If these probes have the ability to mine asteroids and reproduce, then they can grow exponentially and explore the entire galaxy in something like ten to fifty million years. That sounds like a long time, but it's short compared to the lifetime of the galaxy.

But once again, for no good reason other than it lets us continue our train of thought, we will make the simplifying assumption that they use EM radiation and add it to our mental list of necessary coincidences of unknown probability.

If they are not sending us messages but blindly broadcasting into space or just leaking EM radiation from their local equivalent of TV and radio transmission, then it is very unlikely that we would ever hear them unless they were very, very close or we build much bigger telescopes. The signal would be too weak. Our most powerful radio telescope, at Arecibo in Puerto Rico, could only hear such a broadly sent weak signal if it was within about one-third of a light-year from us. But the nearest star to us is more than ten times that distance. For us to receive a message from a distant star, it would almost certainly have to be aimed directly at our cosmic neighborhood, not blindly broadcasted.

Chance That We Are Around at the Same Time (L)

The universe is not just a big place, it is positively ancient. More than 13 billion years of cosmic history is enough time for stars to form, burn, fade, and die several times over. Any of those recent star cycles (once enough heavy elements are made) are good candidates for creating Earthlike planets and conditions hospitable to life. That means that the stretch of time in which an alien race might exist is extremely long. But for us to talk to them, we need to exist around the same time.

How long do technological societies survive? Our limited experience is difficult to extrapolate from, but even human history is filled with cycles of civilization and collapse, on timescales of hundreds of years. Our society is much better equipped to destroy itself than any that came before. Will we be listening for messages in 500 years, or 5,000, or 5 million? Will we even exist?

It is entirely possible that aliens have lived, flourished, sent messages

into space, and then destroyed themselves a million or billion years ago . . . or in the future. For us to talk to aliens, they need to either be very common or survive for a long time.

Imagine if you were still in elementary school and, instead of all the students having recess at the same time, your school randomly assigned recess times to each student. What are the chances that you get to have recess with any of your friends? Or anyone at all? If your recess is five seconds long and there are only two students in your school, you will be playing dodgeball by yourself. If your recess is five hours long or your school has 20 billion students, you are in good shape.

So Where Are They?

Even if we use optimistic values for all of the numbers in the Drake equation and assume that the galaxy is full of long-lived alien technological races, we still have more questions to answer.

Do the aliens even *want* to talk to us? From our perspective, the question might seem absurd: Who would not want to communicate with an alien intelligence? Think of the things we could learn! But that assumes a lot of cultural common ground. We have no idea what these hypothetical aliens might want. Maybe they once communicated with another species and it went badly, and they are taking a ten-thousand-year break from checking their interstellar e-mails and Spacebook updates.

Even in the crazy fortunate scenario that an alien intelligence exists, uses radio to communicate, is nearby, and is sending a message directly to us, could we even tell? While we have radio telescopes listening to the sky, it's not clear that we would know what their message looks like. Sure, we know how *we* would send a message, but in order for the aliens to send us a message that we recognize, we would need to have a whole host of intellectual things in common: symbol-based communication, mathematical encoding systems, similar senses of time, etc. The aliens might think too fast or too slow for us to recognize their message (what if they send one bit every ten years?). The possibility exists that they are sending us messages *right now* but we can't distinguish them from pure noise.

In 1977 a radio telescope in Ohio detected a strange signal. It lasted seventy-two seconds and originated from somewhere in the direction of the Sagittarius constellation. It was so powerful and it varied in strength so much like you would expect a signal from deep space to vary that the scientist on duty that night immediately circled it and wrote "WOW!" on the printout. Unfortunately, the Wow! signal was never heard again (not for lack of listening), and though no convincing Earthbound explanation exists, it can't be unambiguously interpreted as an extraterrestrial message. (That did not prevent scientists from sending a reply in 2012, just in case.)

Even worse are the paranoid scenarios that we can't rule out. Maybe we are surrounded by ancient alien races that avoid contacting us in order

Hello! Anyone out there?
There goes the neighborhood.

to observe our natural progression as if we were in some absurd cosmic zoo. Or perhaps there are many technological races but everyone is listening and nobody is talking out of an abundance of caution and fear of invasion. Or maybe they have already visited but are very stealthy. Given that we know nothing about the hypothetical technology of these hypothetical alien races that hypothetically exist, anything is on the table.

Where Is Everyone?

Why have we not yet found life on other planets? Is it possible that all forms of life are rare or that microbes are common but complex life is rare or that complex life is everywhere but intelligence and civilization are uncommon or that iPad-using tech-savvy aliens are all over the galaxy but are not talking to us or lived and died a million years ago or are talking to us in a way we don't understand?

While it is tantalizing to think of the things we could learn from such an encounter, the dangers of first contact are also real. Consider in human history what happens when a powerful culture meets a weaker culture: it rarely ends well for the more primitive side. Since we don't yet have the capacity to visit other planets or stars, should we be broadcasting our presence and inviting anyone in our galactic neighborhood to drop in and help themselves to the leftover pies in our fridge (or, worse yet, us)?

Could We Learn Physics from Them?

Putting actual physical contact off the table due to the difficulties of manned (or aliened) interstellar travel, what about just talking?

Imagine what such a conversation would be like. Every message would take years (or decades or centuries) to transmit because of the long distances, and in the most optimistic scenario in which their minds work similarly to ours, it would still take several messages to get some basic communication protocols down. The universe's enormous size and slow speed limit mean that any such conversation could take generations. At the rate that our society is changing and our views of science are developing, we might find our own questions silly or poorly chosen by the time we get answers.

Are We Alone?

Perhaps one day you will hold in your hand a Lonely Planet guide to other planets (although by then they might need to change the name to Lonely Galaxy), a book where backpackers can get great recommendations on what to bring to a Hrzxyhpod party on Alpha Centauri or where to get the best tentacle-flavored popsicles on planet Kepler 61b. How big would this book be? Would it be hundreds of pages long, cataloging the zillions of different instances of life that have developed in myriad and strange ways throughout the universe? Or would it be a single solitary page describing only life on Earth?

This remains one of the greatest mysteries of science: How unlikely and unusual is life?

Visit Earth!
Hurry up before the
species that lives
there destroys it!

On one hand, our particular kind of life *seems* very unlikely. Think about all the crazy coincidences that had to happen for you to be here in this moment reading this award-winning physics book.[128] Our star had to be just the right size and temperature, our planet had to be on just the right orbit, and water had to miraculously land here, perhaps flung from the deep reaches of space in the form of comets or ice asteroids. And on this planet, just the right combination of atoms and molecules had to form until one day lightning had to strike to create that first spark of life. How unlikely is it for that spark to have flourished? What incredible odds must it have overcome in an uncaring rocky landscape to grow and one day lead to . . . us? The intricate mechanism of life seems like an unlikely phenomenon to say the least.

But that focuses on our particular type of life. It's true that a long sequence of events had to conspire to specifically produce humans, but in the event that one of those events misfired, maybe another species or another type of life would be in our place. To argue that life is uncommon requires showing that any other sequence of events would have led to a sterile planet. But since we don't know all of the forms that life can take, we cannot make that point.

The reason we don't know how to accurately estimate the conditions that lead to life is that we have only one data sample: us. How do you measure the probability that lightning will strike when you have seen it hit only once? Maybe we are horribly biased by our own experience of how life began on Earth and could be totally blind to maybe millions of other ways in which life can begin. Perhaps *our* life began as an unlikely

128 They give awards to physics books with fart jokes, right?

lightning strike, but maybe there are convenient electrical outlets all over the universe. We have no idea!

And remember, even if life is unlikely, we live in a crazy big universe. It is incredibly vast with billions upon billions of galaxies, each with billions of stars and planets spread out over impossible distances. Whether we are alone in the universe depends on these two factors: is the potential unlikeliness of life overshadowed by the crazy bigness of the universe? If you roll the dice enough times, even the nearly impossible is likely to happen.

One thing is certain though: the truth *is* out there (cue *The X-Files* music). Either there is (or was or will be) life on other planets or there is not. The answer exists totally independent of whether we are here or whether we ask the question or not.

Either answer is mind-blowing and *one of them is true.*

The good news is that we are now getting a real sense of exactly how big the universe is, how it's structured, and how many planets are in it. For the first time in the history of life on this planet, we have opened our eyes and spread the reach of our knowledge almost as far as is possible.

Perhaps we are alone in the universe and human beings are the only beacon of self-awareness that this vast cosmos has and ever will know.

Or perhaps the universe is teeming with life at every corner, and we are just one of the millions of different ways in which molecules can arrange themselves in self-replicating, consciousness-baring, eyeball-eating ways.

Or maybe the answer is somewhere in between, and life is rare but not that rare. Perhaps there will be only a few outposts of life in the history of the universe and they will never talk or know about one another because of the enormous scales of space and time.

In all cases, we should never forget: life exists, and we are the proof of that.

A Conclusion of Sorts

The Ultimate Mystery

And so we come to the end.

If you bought, borrowed, or stole this book because you wanted answers to the biggest questions in the universe, then maybe this wasn't the right book for you.[129] This book is not so much about answers but about questions.

Over the last seventeen chapters, you've learned that we still have much to learn about *a lot* of things. A lot of pretty *big* things. Knowing that we don't know what 95 percent of the universe is made of or that there are strange things out there that we understand very little about

129 We realize it's a tad late for this warning.

(antimatter, cosmic rays, the speed limit of the universe) might make you a little distressed. Who wouldn't be after finding out that you are surrounded by an unknown substance called dark matter and are being pulled on by something called dark energy *at this very moment*? It's enough to make anyone a little nervous about stepping outside their front door.

But we hope that you've also learned the main lesson of this book: that we should be *excited* about all the things we don't know. The fact that we still don't know so many fundamental truths about the universe means there are still incredible discoveries ahead of us. Who knows what amazing insights we will find or what mind-blowing technologies we will develop along the way? The age of human exploration and discovery is far from over.

If you've truly taken this lesson to heart, then perhaps you are ready for us to discuss one last mystery in this book. And it starts with a question so deep and profound that many might call it the Ultimate Mystery:

Why does the universe exist and why is it the way it is?

At this point, some of you might be a little concerned that we are raising this question. After all, another one of the big lessons of this book is to be mindful about the *bounds of science*. Of all the questions you can ask, there are some that are within the scope of science because their answers are testable. Other questions, whose answers cannot be experimentally tested, may be deep and fascinating but are beyond the scope of science and belong more to the realm of philosophy. Asking why does the universe exist sounds dangerously close to the type of question that belongs in the philosophical category.

Why? Because when you ask this question, what you are really looking for is an explanation based on some fundamental law or fact about the universe that shows the universe *had* to exist and that it couldn't have been made any other way (and still be consistent). If it could have been made another way (or not at all), then another question pops up: Why is the universe *this* way and not *that* way?

But even if you find such an explanation and you discover that there are fundamental laws that couldn't have been put together any other way (i.e., with no arbitrary or random parameters), then yet *more* questions pop up:

> *Why do fundamental laws exist? And why does the universe follow them?*

As you can see, these are all tricky questions even for people in philosophy, and it's clear that the answers may be outside the scope of science.

In fact, it may be that *many* of the deep mysteries we explained in this book are also beyond the scope of scientific inquiry. Does this mean that we will *never* find the answers to these questions?

Not necessarily!

The Testable Universe

It's possible that there are questions for which we will never find the answers, but there are also questions that have moved from philosophy to science. As we expand our ability to look far into the universe and deep within particles, we also expand the number of things we can test with science. This grows what we call the testable universe.

You might recall the concept of the observable universe from earlier in this book. This is the fraction of the universe that we can actually see today because enough time has passed since the beginning of the universe for light from this fraction to reach us. Everything outside of that fraction is invisible to us because the light from it has not yet been able to reach us.

Similarly, the testable universe is the fraction of the universe that we can confirm and know about using science. It doesn't include just the outward bounds of our vision (how far away into space we can see). It also includes the inward bounds (the smallest bits of space and matter that we can see). It includes the limits of how finely and how accurately we can discern things at the smallest and largest scales, and it includes the limits of our theories, mathematics, and capacity for understanding.[130]

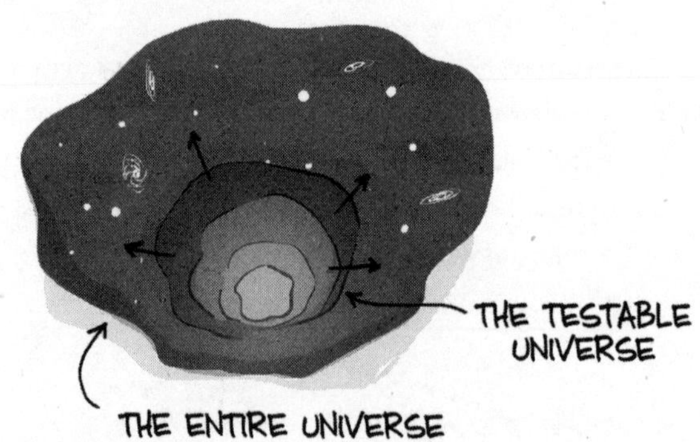

Like the observable universe, it is likely (if not obvious) that the testable universe is much smaller than the complete universe. That means that a whole lot is still beyond our grasp. But here is the exciting part about all of this: even though there are still a lot of questions that are outside of the bounds of science, science is *always growing.*

Like the observable universe, the testable universe is expanding. Every time we develop new technologies and new tools to probe reality, the testable universe grows. Our capacity for understanding the world around us and answering all of the known unknowns in the universe expands

130 This last one is somewhat terrifying: What if the universe makes perfect sense and can be described by a beautiful mathematical theory that is beyond the capacity of our brains to grasp?

each year. In fact, what is amazing is that the growth of the testable universe is *accelerating*.

A few hundred years ago, when science was in its infancy, the testable universe was still pretty small and growing slowly. Our technology and our capacity to model and understand nature were pretty limited during the first few decades of scientific inquiry.

Then a little more than one hundred years ago, as technological progress gave us new tools to explore our environment, the testable universe began to grow rapidly. We could now ask—and answer!—questions about quantum physics, the formation of the universe, and the nature of matter that had previously been left to philosophers.

SCIENCE HITS PUBERTY

It is fair to say today that the testable universe is undergoing its own version of cosmic inflation: an expansion beyond anything we've seen before. From just a hundred or so years ago, we can now peer deep into the Big Bang and perhaps out into the edges of the cosmos. We can speculate and potentially verify whether space itself is infinite or whether it curves around like a potato. We can look deep within protons and accelerate matter to 99.999999 percent of the speed of light. We've even begun to send unmanned spaceships beyond our solar system and land probes on comets.

What does this mean for questions such as "Why does the universe exist?" which today seem to us hopelessly outside of the testable universe?

We should look to recent history and be encouraged by the rapid inflation of our knowledge. The scientific tools and techniques being developed today and in the future will keep increasing the number of things we can study and the number of questions that can have firm and testable answers.

Will we one day be able to answer deep questions like this about the universe?

We have no idea.

But it will certainly be an exciting ride.

STAY TUNED FOR OUR SEQUEL,
WE HAVE SOME IDEAS.

Acknowledgments

We are grateful for invaluable scientific insight and fact-checking by James Bullock, Manoj Kaplinghat, Tim Tait, Jonathan Feng, Michael Cooper, Jeffrey Streets, Kyle Cranmer, Jahred Adelman, and Flip Tanedo.

Many thanks to readers of early versions who provided feedback: Dan Gross, Max Gross, Carla Wilson, Kim Dittmar, Aviva Whiteson, Katrine Whiteson, Silas Whiteson, Hazel Whiteson, Suelika Chial, Tony Hu, and Winston and Cecilia Cham.

Special thanks to our editor, Courtney Young, for her faith in this project and her steady guidance, and to Seth Fishman for helping us find the right home for this book. Thanks to the whole team at the Gernert Company including Rebecca Gardner, Will Roberts, Ellen Goodson, and Jack Gernert. And many thanks to everyone at Riverhead Books who contributed their time and talent to the making and release of this book, including Kevin Murphy, Katie Freeman, Mary Stone, Jessica Miltenberger, Helen Yentus, and Linda Korn.

We also want to thank the many people who have been following our work online for years. You inspire us to keep doing interesting things.

Finally, we thank the many, many scientists, engineers, and researchers who work to push the boundaries of our knowledge. Here's to your ideas.

Bibliography

How Do You Know This Stuff? And Where Can I Learn More?

Chapters 1 and 2

The fraction of dark matter and dark energy mentioned here comes from the 2013 measurement by the *Planck* Collaboration: https://arxiv.org/abs/1303.5062. Updated measurements exist, but the qualitative story has not changed.

Galactic rotation curves were first studied by Vera Rubin and Kent Ford in the 1960s and '70s. Rubin, Vera, Norbert Thonnard, and W. Kent Ford Jr. 1980. *The Astrophysical Journal* 238: 471–87.

Gravitational lensing is really two different approaches. Strong lensing shows the dramatic distortion of a single galaxy (e.g., https://arxiv.org/abs/astro-ph/9801158), and weak lensing measures minute effects on many galaxies on a statistical basis (e.g., https://arxiv.org/abs/astro-ph/0307212).

The galactic collision mentioned is the Bullet Cluster (https://arxiv.org/abs/astro-ph/0608407). We learn from that collision that dark matter does not have strong self-interactions (https://arxiv.org/abs/astro-ph/0309303).

For a review of the current knowledge of dark matter and searches for WIMPs: http://arxiv.org/abs/1401.0216.

Chapter 3

Type Ia supernovae were observed by the High-Z Supernova Search Team (https://arxiv.org/abs/astro-ph/9805201) and the Supernova Cosmology Project (https://arxiv.org/abs/astro-ph/9812133). These supernovae do not all have the same peak brightness, but they have a characteristic light curve, the amount of light they emit as a function of time, that can be calibrated (Phillips, Mark M. 1993. *The Astrophysical Journal* 413, no. 2: L105–8) so that these supernovae provide a measure of distance.

Chapter 4

Many details about our current understanding of particles can be found at the Particle Data Group's website: http://pdg.lbl.gov.

Chapter 5

$N = 10^{23}$ or so is when the energy of the bonds starts to match the energy of the llama pieces because that's roughly the number of atoms in a macroscopic object (the Avogadro constant).

Experimental observations of binding energy affecting mass include radioactive decay processes—neutron beta decay, for example. A neutron of mass 939.57 MeV decays into a proton of mass 938.28 MeV, an electron of 0.511 MeV, and a neutrino of negligible mass. The disappearing mass (939.57 – [938.28 + 0.511] = 0.78 MeV) is a result of the lower energy of the proton's bonds and is converted into kinetic energy of the proton, electron, and neutrino. An opposite example is the O_2 molecule, which has *less* mass than two oxygen atoms because the two atoms attract each other and the formation of the O_2 molecule *releases* energy.

The percentage of 0.005 comes from the fact that the average binding energy per nucleon is a few MeV (one to nine typically) and the mass of nucleons is nearly 1,000 MeV.

The masses of the up and down quarks are <5 MeV, and the mass of the proton and nucleon are about 1,000 MeV, so the total masses of the quarks inside the nucleons is about 15/1000 or about 1 percent.

The top quark has a mass of around 170,000 MeV, and the up quark of 2.3 MeV, for a ratio of about 1:75,000.

A technical description of how the Higgs field works and how it solves the problem of the masses of the W and Z bosons can be found in http://arxiv.org/abs/0910.5095 or in more detail in the authors' video: https://vimeo.com/41038445.

Chapter 6

There are several ways to compare the strength of gravity to other forces.

First, we can compare the gravitational coupling constant ($\alpha_g = Gm_e^2/\hbar(c) = 1.7518 \times 10^{-45}$) to the electromagnetic coupling constant (also called the fine-structure constant) of $1/137 = 7 \times 10^{-3}$. That's a ratio of 10^{-42}.

But the force felt on objects due to gravity and electromagnetic force also depends on the mass and the charge. For example, if you compare the gravitational versus electromagnetic forces on two protons (charge = 1, mass = 1000 MeV):

$F_g = G(m_p m_p/r^2)$

$F_{em} = k_e(q_p q_p/r^2)$

Then $F_g/F_{em} = G(m_p m_p)/k_e(q_p q_p) = [G(m_p)^2] / [k_e(q_p)^2] = [6.674 \times 10^{-11}\ Nm^2/kg^2\ (1.67 \times 10^{-27}\ kg)^2] / [8.99 \times 10^9\ Nm^2/C^2\ (1.6 \times 10^{-19}\ C)^2]$

$= 8 \times 10^{-37}$, *which is close to* 1×10^{-36}

A gravitational wave will distort space by a tiny amount. The first detection by LIGO saw one with distortion of about 1×10^{-21} (Fig 1 of https://arxiv.org/abs/1602.03837).

Chapter 7

Space is flat to within 0.4 percent, according to the WMAP 2013 measurement of the cosmic microwave background (http://map.gsfc.nasa.gov/universe/uni_shape.html) and studies of large triangles (https://arxiv.org/abs/astro-ph/0004404).

Distances of 10^{-35} meters refer to the Planck length scale: $\sqrt{(\hbar G/c^3)} = 1.616 \times 10^{-35}$ meters.

Chapter 8

For further reading on the arrow of time, we recommend the excellent book *From Eternity to Here* by Sean Carroll.

Chapter 9

Solar neutrino flux is about 7×10^{10} particles per cm^2 per second (from Claus Grupen's *Astroparticle Physics*, page 95).

See the notes for chapter 6 on the weakness of gravity for a discussion of the 10^{-42} factor between gravity and the electromagnetic force.

The footnote on quantum mechanics and its view of time refers to the uncertainty principle, which can relate uncertainty in energy to uncertainty in time.

Chapter 10

For a good review of relativity, see *Modern Physics for Scientists and Engineers* by John R. Taylor, Chris D. Zafiratos, and Michael A. Dubson.

The speed of light is 299,792,458 meters per second. That is an exact number since it is used to define the length of a meter.

The limits of human tolerance to g-forces have been studied in the context of fighter pilots (see *Medical Aspects of Harsh Environments*, volume 2, chapter 33, by Ulf Balldin).

Acceleration at 3 g (~30 m/s^2) would take 10 million seconds (one-third of a year) to reach the speed of light, but note that maintaining that acceleration would require increasing amounts of energy.

The nearest star is Proxima Centauri, 4.2 light-years = 4.0×10^{16} meters.

Chapter 11

A review of cosmic rays and their detection mechanisms is in *Extensive Air Showers* by Peter Grieder.

The slowing down of ultra-high-energy cosmic rays due to their interaction with photons from the early universe is called the GZK (Greisen-Zatsepin-Kuzmin) effect.

Note that many of the numbers in this chapter are approximate as the rate of particles at very high energies has large uncertainties, but the qualitative story is unaffected.

Chapter 12

CERN can make 10 million antiprotons per minute (See "Cold Antihydrogen: A New Frontier in Fundamental Physics" by Niels Madsen, published in *Philosophical Transactions of the Royal Society* in 2010).

A gram of antimatter plus a gram of matter would release 2 grams × c^2 energy = $(2 \times 10^{-3}$ kg$)(3 \times 10^8$ $m/s^2)^2 = 1.8 \times 10^{14}$ J = 43 kilotons.

The search for antimatter galaxies is described here: http://arxiv.org/abs/0808.1122.

The ALPHA experiment at CERN produces and analyzes antihydrogen. See https://home.cern/about/experiments/alpha.

Chapter 14

The age of the universe is 13.6 (13.8) billion years, according to 2013 (2015) *Planck* data.

The accidental discovery of cosmic microwave background by Arno Penzias and Robert Wilson in 1964 won them the 1978 Nobel Prize in Physics. The time that the universe became transparent is dated to 380,000 years after the Big Bang according to the 2013 *Planck* data (https://www.mpg.de/7044245).

There are many theories of inflation; here we use rough numbers that are characteristic: 10^{25}-fold expansion for a short period starting the first 10^{-30} seconds after the Big Bang.

Chapter 15

The number of stars in the Milky Way is not firmly known. Estimates vary from 100 billion to up to a trillion (http://www.huffingtonpost.com/dr-sten-odenwald/number-of-stars-in-the-milky-way_b_4976030.html).

The number of galaxies in the observable universe is also not firmly known. Estimates vary from 100 to 200 billion (http://www.space.com/25303-how-many-galaxies-are-in-the-universe.html) up to 2 trillion (https://arxiv.org/abs/1607.03909).

The mass estimate of our supercluster is 10^{15} times the mass of our Sun (https://arxiv.org/abs/0706.1122).

Simulations show that galaxy formation depends on the presence of dark matter (http://arxiv.org/abs/astro-ph/0512234).

The size of the observable universe is estimated to be 14.26 gigaparsecs, or 46.5 billion light-years, 4.40×10^{26} meters in any direction (https://arxiv.org/abs/astro-ph/0310571).

The estimate of the number of particles in the universe is very rough and mostly comes from the estimate of the number of stars and the ratio of dark matter to normal matter; since the mass of dark matter is unknown, there is a large uncertainty. See: http://www.universetoday.com/36302/atoms-in-the-universe/.

Chapter 16

The radius of a proton is about 10^{-16} meters, but the definition is a little philosophical.

The LHC has collision energies of about 10 TeV, which is 10^{13} eV corresponding to 10^{-20} meters.

Chapter 17

The estimate of the fraction of stars with Earthlike planets comes from *Kepler* data (http://arxiv.org/abs/1301.0842).

The *Meteoritical Bulletin* database currently lists 177 meteorites cataloged as having Martian origins (http://www.lpi.usra.edu/meteor/index.php).

Recommended Reading

Our Mathematical Universe by Max Tegmark, published by Knopf in 2014.

From Eternity to Here by Sean Carroll, published by Dutton in 2010.

Seven Brief Lessons on Physics, by Carlo Rovelli, published by Riverhead in 2015.

Index

C

D

M

N

O

P

T